Ecology, Artificial Intelligence, and Virtual Reality

Environment and Society

Series Editor: Douglas Vakoch

As scholars examine the environmental challenges facing humanity, they increasingly recognize that solutions require a focus on the human causes and consequences of these threats, and not merely a focus on the scientific and technical issues. To meet this need, the Environment and Society series explores a broad range of topics in environmental studies from the perspectives of the social sciences and humanities. Books in this series help the reader understand contemporary environmental concerns, while offering concrete steps to address these problems.

Books in this series include both monographs and edited volumes that are grounded in the realities of ecological issues identified by the natural sciences. Our authors and contributors come from disciplines including but not limited to anthropology, architecture, area studies, communication studies, economics, ethics, gender studies, geography, history, law, pedagogy, philosophy, political science, psychology, religious studies, sociology, and theology. To foster a constructive dialogue between these researchers and environmental scientists, the Environment and Society series publishes work that is relevant to those engaged in environmental studies, while also being of interest to scholars from the author's primary discipline.

Titles in the Series

Living Deep Ecology: A Bioregional Journey, by Bill Devall and edited and introduction by Sing C. Chew

Environmental and Animal Abuse Denial: Averting Our Gaze, edited by Tomaž Grušovnik, Reingard Spannring, and Karen Lykke Syse

Conservation, Sustainability, and Environmental Justice in India, edited by Alok Gupta

Secular Discourse on Sin in the Anthropocene: What's Wrong with the World? by Ernst M. Conradie

Environment, Social Justice, and the Media in the Age of Anthropocene, edited by Elizabeth G. Dobbins, Maria Lucia Piga, and Luigi Manca

Global Perspectives on Eco-Aesthetics and Eco-Ethics: A Green Critique, edited by Krishanu Maiti and Soumyadeep Chakraborty

Australian Wetland Cultures: Swamp Thinking, edited by John Charles Ryan and Li Chen

Motor Vehicles, the Environment, and the Human Condition: Driving to Extinction, by Hans A. Baer

Ecology, Artificial Intelligence, and Virtual Reality: Life in the Digital Dark Ages, by Sing C. Chew

Ecology, Artificial Intelligence, and Virtual Reality

Life in the Digital Dark Ages

Sing C. Chew

LEXINGTON BOOKS
Lanham • Boulder • New York • London

Published by Lexington Books
An imprint of The Rowman & Littlefield Publishing Group, Inc.
4501 Forbes Boulevard, Suite 200, Lanham, Maryland 20706
www.rowman.com

6 Tinworth Street, London SE11 5AL, United Kingdom

British Library Cataloguing in Publication Information Available

Library of Congress Cataloging-in-Publication Data

Names: Chew, Sing C., author.
Title: Ecology, artificial intelligence, and virtual reality : life in the digital dark ages / Sing C. Chew.
Description: Lanham, Maryland : Lexington Books, [2021] | Series: Environment and society | Includes bibliographical references and index.
Identifiers: LCCN 2020050849 (print) | LCCN 2020050850 (ebook) | ISBN 9781793641502 (cloth) | ISBN 9781793641519 (epub)
Subjects: LCSH: Environmental protection—Data processing. | Sustainable development—Data processing. | Global environmental change. | Technology—Social aspects. | Social change.
Classification: LCC TD170 .C496 2021 (print) | LCC TD170 (ebook) | DDC 304.20285/63—dc23
LC record available at https://lccn.loc.gov/2020050849
LC ebook record available at https://lccn.loc.gov/2020050850

For Steve Newman and Dan Sarabia

Contents

Preface

This book is about system transition and transformation. It is my belief and shared by others as well, such as the late Terry Hopkins and Immanuel Wallerstein (1996, 1998), that for some time now the world system is in a state of crisis, and that the structures and processes that maintain the reproduction of the world system no longer function effectively to maintain equilibrium. Unlike phases of economic downturns that the world system goes through periodically as a consequence of the accumulation of capital dynamics, what we have been experiencing reflects more than this; they are signs of structural breakdowns and the emergence of new structures and processes that in the long-run will have transformative effects on the system. Over *la longue durée*, I also suspect that we all will increasingly realize that these crisis conditions are insurmountable, and that a system transition is taking place. Given such a *conjoncture*, this book examines further the structures and processes that were identified in my book *Ecological Futures* more than a decade ago, and suggests pragmatic technological and ideological solutions that are increasingly available to address the global environmental issues that we face. The solutions we are offering can, perhaps, allow natural and social evolution to continue even though our technologies and our materialistic consumption practices have impacted on the trajectory of natural evolution.

Some readers might categorize what I am proposing as a "technological solutionism" to meet present and future environmental challenges. If the fear is that these technologies will be detrimental to our social and natural systems, my response is that they are already in our midst whether we like it or not! In fact, many have embraced warmly the technologies that I have identified. I believe, at this point in world history, we will need to adapt and be pragmatic to ensure that natural and social evolution can and will continue. In light of the multiplicity of postmodern narratives and tropes on display, and in practice these days, my stress on the importance of social and natural evolution continuing is, for me, the only grand narrative left for us to be concerned with under current circumstances.

Of course, there will be carpers to the suggestions made in this book. We are all aware that there are often, at least, two opinions especially when technological innovations are considered as a response to socioeconomic and ecological issues confronting us (Dauvergne 2020a). The technological solutionism side tends to see the rosy aspects of what the utilization of technology can do to deal with the global environmental problems and socioeconomic issues we face. Included in this are those corporations and individuals with the power who have also amassed the monetary profits from digitalization and control for what happens. On the other side of the divide are those that have a 'natural' adverse reaction to technology and who also fear and worry that the nature of this type of technology can be utilized for social control, and in the long run impact on our freedom and privacy (Dauvergne 2020a). These concerns are addressed as well in this book.

As the new millennium progresses, given what I identified in terms of the global structures, processes, and phases underlying the decline of the Bronze Age in my trilogy on the history of world ecological degradation have reappeared, albeit with different particularities (Chew 2001, 2006, 2007). With the end of the Bronze Age we have the transition to the Iron Age. It looks like now we are also experiencing the end of the Iron Age and the transitioning to the Silicon Age. At a lightning pace, transformations in this Age of Silicon are occurring, especially in the field of Artificial Intelligence, Virtual and Augmented Realities, Robotics, Cloud Computing, and Machine Learning, needs to be examined in more detail, and how they can assist us in dealing with natural resource scarcities, biodiversity crisis, climatological changes, disease pandemics, water scarcities, and food production. This book identifies the technological advances and innovations that have been made in our increasingly digitalized world, and proposes how these new technologies can be utilized to address the environmental issues we face.

Material reducing processes via the concepts of dematerialization and miniaturization need to be encouraged and promoted as avenues to reduce natural material inputs in the socioeconomic production processes. Even political processes are increasingly using social media to communicate societal policies and goals that are now part of the political landscape. Clearly, the costs of reproducing our varied lives as we know it has changed, and the degree of savings in terms of natural resources need to be worked out carefully. I have explored this theme, but more needs to be done.

Commodity exchange has been the basis of human civilization since the dawn of time. Over the last five thousand years of world history, the basis of exchange has been determined by value of the commodity. The valuation process underlying the exchange has been considered exploitative, and hence, unequal by some scholars. Various models have been proffered. Within the ambit of this book, I focus on the valuation process of symbols and signs that have no materiality. To me, we have arrived at a qualitative moment in world history whereby non-material exchanges permeate our lifestyles significantly. Such a shift suggests that we are in the midst of major transformations in our consumption habits and practices, and that new horizons of distribution of surpluses between regions and within regions of the world system are in the process of forming. These are interesting times whereby individuals and nations in what is now considered the Global South will have opportunities to leapfrog and participate more in the accumulation of capital that would not have been possible in the past. The globalized interconnections have opened up possibilities.

There are some aspects of the reproduction of life that are material based. The most important is the production and consumption of food. Here, we are also seeing the various opportunities that the new digitalized technologies can contribute to the consumption needs.

Like the end of the Bronze Age, we are also witnessing social movements with different ideologies that are quite different from the technologistic ideas that pervade a digitalized world. These movements offer a set of ideas that underline aspects of lifestyles that some have categorized and labeled as traditionally rebellious in demanding quality over quantity, natural versus genetically manipulated, and healthy over disease-prone. How will this ideological transformative process end? If what happened at the end of the Bronze Age can be our guide, over time, hybridization of ideas and processes will determine the trending direction.

Crisis does force change. The change can be an opportunity for reconsideration of ongoing goals and force us to make choices that were never considered in the past. Take the case of the current disease pandemic. It is not new! It has happened countless times in the past during the Dark Ages. What the current circumstance has placed us in is an overwhelming reliance on digitalized technologies to reproduce our lifestyles, and a non-questioning of scientific knowledge to aid us through these turbulent times. Such developments further cement our transition to the Silicon Age, and wed us further to the digital cyber-world. In the process, Digitalization and Science have become the ideology of the Silicon Age!

Acknowledgements

It is always a pleasure to acknowledge friends and colleagues who have provided input and comments that, in my view, have made what has been discussed in this book clearer. Dan Sarabia, Bill Thompson, Tom Hall, Mitch Allen, Amanda Bertana, Bob Denemark, Al Bergesen, Mark Bauermeister, Pat Lauderdale, Charles Clark, Peter Jacques, Mike Levy, and Gary Strokos provided valuable insights. Steve Newman has influenced my thinking on Artificial Intelligence, and its place in our global futures. Also, much thanks goes to my editor, Mike Gibson, who has been extremely supportive of this project, and has made the completion of this book so satisfying.

'Place' plays a major part in determining one's bearing and senses. One's relationship to the landscape, to the ever-changing rhythm and smell of the surroundings, and to the conversations one has with other humans conduce us to be more in rhythm to the layout and memories, and it enables us to be able to contrast such life-experiences to other places we have been. Because of certain circumstances, my wife, Elizabeth, and I including our dog, Elsa, had to live for a year in the city of Fremont located in East Bay, "Silicon Valley," California. Our everyday rush hour drive from Fremont to Stanford University located in Palo Alto, depending on traffic conditions on interstate 101 and 280 and the number of "Google buses," meant passing Redwood City, Menlo Park, Cupertino, Mountain View, Sunnyvale, and Santa Clara. All these cities housed most of the headquarters of the major high tech companies of the global world. Every day, seeing tourists having their pictures taken in front of the "Facebook Thumbs Up Sign" wall located next to the Willow Road entrance of Facebook's HQ campus on Bay Front Expressway brought back memories of tourists having their pictures taken in the front doorsteps of the Notre Dame Cathedral in Paris. It looks like the campus buildings of the high tech companies have replaced the cathedrals of Europe!

Furthermore, my conversations with high-tech contract employees and consultants who were based in/or from different parts of the world, over dinner and breakfasts in the extended-stay hotel we were all living in, also made me realize how transitory it was in terms of project length, directions, and the global nature of the information technology business. At the end, I was in an informatics mindset that led to the tone and directions taken in this book. In this light, Place has shaped the writing of this book, and the reader will sense, and might even object to, my overall enthusiasm and the promise of the different technologies discussed in this book.

Sing C. Chew, McKinleyville and Fremont, California

Chapter 1

Dark Ages and Contours of System Transition

INTRODUCTION

The word "crisis" comes easily to our lips these days. To most of us, regardless of whether the crisis is political, economic, social, climatological, or ecological in nature, the world seems quite chaotic and upsetting. Chaos is antithetical to the modern progressive thinking; to our modernistic *mentalité* this chaotic and dangerous state of our world is not what the years of human progress are supposed to achieve.

Modernity has conduced us to accept and take progressive actions in order that the good and just society can unfold. Therefore, political, economic, and social actions and policies are pursued to minimize the chaotic conditions of our world so that the growth and progress that we all yearn for can continue unabated. Along with this, the *mentalité* of modernity has also determined and shaped the way we evaluate, rationalize, and plan what we do to meet our life challenges. What it entails, therefore, is the continued push for the "rational" expert/scientistic way to achieve the goals of the cultural assets of our society such as political governance, socioeconomic transformations, and knowledge generation.

The crisis conditions that we are experiencing these days suggest that perhaps the system is experiencing problems in maintaining the reproductive capacity for the system to function effectively in an equilibrium fashion. We must also note that crisis and chaos are not just depressed conditions experienced by modern societies. In the past, before the arrival of modernity and its associated characteristics, hierarchical societies in order to meet their reproductive needs also generated crisis conditions. Such conditions throughout world history for the last five thousand years have engendered socioeconomic setbacks, cultural upheavals, and environmental destruction and pollution;

and these prolonged periods of crisis have been termed as the Dark Ages (Chew 2001, 2007, 2008; Snodgrass 1971; Braudel 2001; Desborough 1972).

Accordingly, what past hierarchical societies have experienced can, perhaps, also provide insights for us on how to resolve our current crisis conditions. If this is the case, we should be alerted to the fact that Dark Age conditions in our historical past that have occurred, and that cannot be resolved, have also produced a social reorganization and a system transition. Such occasions are rare. Over the last five thousand years of world history, we know of at least one social system transition, and that was at the end of the Bronze Age of human history (Chew 2007). The Bronze Age social system transition generated the emergence of different socioeconomic and political structures producing socioeconomic and political practices that were qualitatively different. That Bronze Age system transition was depicted by the process of destructuration and restructuration of our social fabric of life (Chew 2008).

If the above is what happened in our social evolutionary process, we should also ask, especially in this present era, what was the condition of Nature then during such a period of system transition since human societies are intimately bound to Nature in their reproduction of their lifestyles? Interestingly and historically, instead of the chaos we witnessed in the socioeconomic and political realm during the Dark Ages, we find a rejuvenated landscape and a revived Nature (Chew 2001, 2006, 2007). Nature was restored because there was less resource demand due to the major slowdown in socioeconomic activity that occurred during a Dark Age. In this vein, such a slowdown in socioeconomic activities (as a result of the crisis) provided a "time respite" for Nature to recover (Chew 2006). In this light, socioeconomic and political chaos along with slowdowns during Dark Ages generated the opportunity for Nature to rest. From the past, what was shown was that the system crisis leading to the period of system transition was good for Nature. The air turned clear and the rivers flowed!

From what we have witnessed in the last five thousand years of world history, and in light of the contemporary global chaotic/crisis conditions, the world system is again in a phase of system transition. Such a view is also shared by scholars, such as Terry Hopkins and Immanuel Wallerstein, who recognized that the capitalist world system is in the phase of transition because the mechanisms of the system that maintain equilibrium no longer function effectively for system reproduction (1996, 1998).

Given the above, the contemporary global chaos and system crisis have many thematics, trends, and tendencies (see, for example, Denemark et al. 2000; Chase-Dunn and Anderson 2005; Chew 2008). Instead of trying to cover broadly every trend and thematic on this current phase of system transition like what had occurred in the past during the Bronze Age as I have done in my book, *Ecological Futures*, I will, instead focus on a selected set of trends that, in my view, has considerable impact on the ongoing chaos and threats to the reproduction of life. In particular, therefore, the focus will be on how do we adapt and evolve to the impending threats that we will face such as scarcity of natural resources, global warming, food accessibility to meet population needs, and the issue of living in a built environment. To meet these challenges, an in-depth examination of the technological advances and innovations of the cyber-world will be undertaken to assess whether these technological developments can be harnessed so that social and natural evolution can continue.

THE NATURE OF DARK AGES AND GLOBALIZATION

For the disciplines of sociology and political science, there are basically two narratives in trying to explain the process of globalization that the world has experienced. The first, which is the orthodox way of explaining globalization, argues that the process is a stage attained following the accumulation of wealth at the national state level in one geographic region of the world, for example, in Western Europe. This process of globalization began at the nation-state level, and over time, this pool of capital accumulated in separate developed nation-states was followed with capital exports via multinational corporate business activities from these nations to other parts of the world. Over the course of history, from the A.D. nineteenth century onwards, this progression of multinational economic and political activities managed to cover the globe extensively.

The other narrative explains globalization differently. To the proponents of this second narrative, globalization is an evolving interconnected process that has occurred throughout world history, and that it is not a stage reached but one that started at perhaps different time periods. In this regard, the world was linked globally even at the start of world history as early as five thousand years ago in a connected process of trade and economic exchanges. Different time phasings have been identified in terms of the formation of an interconnected globalized world with one view envisaging the globalization process starting from five thousand years ago to another position identifying this process as beginning in the sixteenth century AD (Wallerstein 1974; Frank 1998, 2015; Modelski et al. 1999; Thompson 2000; Chase-Dunn and Hall 1997; Chew 2001, 2007).

If we consider the above latter position that globalization is an evolving process, we can posit the perspective that there is the emergence of a structure that can be categorized as a world economy/system. One proponent of this narrative, Immanuel Wallerstein (1974), however, insists that this world economy is distinguished necessarily by a distinct mode of production (capitalism) covering a distinct period in time from the sixteenth century onwards. Whereas for the other proponents of this perspective, their only distinction is that this world economy/system covers a geographic space with multiple cores/regions linked at a minimum by a trading system, time phasings notwithstanding (Frank 1998, 2015; Modelski et al. 1999; Thompson 2000; Chase-Dunn and Hall 1997; Chew 2001, 2007). It is an evolving global economy "of the world." Depending on the temporal sequence, it is an economy of the world encompassing different chiefdoms, kingdoms, civilizations, empires, and states in a global division of labor, technology, and knowledge circumscribed by different cultural patterns (Modelski et al. 1999; Chew 2001, 2007, 2018; Frank 1998, 2015; Chase-Dunn and Hall 1997; Thompson 2000).

Regardless of time or geographic space, this historical globalization process is punctuated by system crises that impact on the regions of the world. These systemic crisis phases I have termed as the recurring Dark Ages (Chew 2001, 2007, 2008). The time phases of these recurring Dark Ages are as follows: 2200 B.C.–1700 B.C.; 1200 B.C.–700 B.C.; and A.D. 400–A.D. 800/900. The extent of the spread of such crisis conditions and its geographic coverage is determined by the systemic connections of this world economy that continues to evolve and transform to overcome these systemic crisis phases or Dark Ages.

We should note as well, that the Dark Ages are not only distinguished by slowdowns and downturns in socioeconomic conditions and increases in political rivalry and regional wars; depending on the state of the environment then there is perhaps reduced environmental degradation and species extinction. Other distinguishing features of Dark Ages are large-scale human migrations, disease pandemics, climate changes appearing as droughts, floods, and temperature increases. To this extent, we should also underscore that climate change is one of the main factors in the precipitation of Dark Ages (Chew 2008; Thompson 2000).

When it comes to Nature, these Dark Ages (we must realize) were by no means continuously exhibiting increasing ecological degradation. This is because during these Dark Ages, the decline in socioeconomic activity provides Nature the opportunity to recover. It is important further to record that these Dark Age periods appear when the system has reached the nadir of a crisis point whereby system maintenance and continuance is stalled. If some structural resolutions do not emerge in order for the system to persist to evolve ecologically, socioeconomically and politically, system transition becomes the order of the day.

If Dark Ages are prolonged crisis periods, such crises provide opportunities. In other words, crisis conditions provide the opportunities for the resolution of contradictions that have developed to such a state that inhibits the reproduction of the world system. It leads to pathways and processes that would mean system reorganization and transition. If reorganization does not occur, system collapse usually follows. This has been documented in many places historically (see, for example, Kristiansen 1993, 1998; Chew 1997, 1999, 2001). Furthermore, what needs underscoring is that the limits of the natural system become also the limits of the socioeconomic processes of the world system; and the interplay between ecological limits and the dynamics of the social system define the historical trajectories of human history (Kristiansen 1998; Chew 2001, 2002).

The Long-Term Past

Following Dark Ages or crisis periods are periods of growth and with the historical globalization process we usually find an increase in density of trade connections. If we trace the course of world history starting five thousand years ago, we can distinguish the pulsations of this timing of crises followed by growth phases (Chew 2007). Let us follow historically these pulsations of crisis and growth.

Reflecting on the long-term historical past, in terms of the geographic coverage of the world system, we find that by 3000 BC, the contours of a "regional" world economy encompassing the Eurasian region of Mesopotamia, the Arabian Peninsula, Levant, Anatolia, Iran, the Indus Valley, and Egypt can be demarcated (Chew 2001, 2007; Frank 1998, 2015; Chase-Dunn and Hall 1997; Thompson 2000). This expansion of geographic growth produced by the turn of A.D. 1 or even earlier, an interconnected world system encompassing Europe, East Africa, and Asia (South, Southeast, and East) (Beaujard 2010; Chew 2001, 2007, 2018; Frank 1998, 2015). In world history, we can conceive this as the first Eurasian world economy because the only major region that has not been connected at this point in history is the Americas (Chew 2018).

On another note, if Nature is brought back into the picture in the growth and expansion of the world system, barriers to the reproduction of the world system are formed when human societies induced changes to the ecology and climate. These changes are caused mainly by human-generated social organizational factors such as urbanization, accumulation, wars, technological innovations, and population that impact on social and natural system reproduction. In terms of these social organizational factors, it is important to realize that urbanization and population have tremendous impacts on the landscape and ecology, thus the reproduction of culture and life as we have witnessed to date.

Notwithstanding these above factors, natural disturbances such as earthquakes and volcanic eruptions can also condition the reproduction and evolution of the world system, and thus work independently. Again to repeat, climate as well can precipitate and affect the crisis independently, and also dependently. It can conduce the rise of disease pandemics and other socioeconomic and political outcomes.

As we have suggested in the previous pages, through the course of human history, system crises have appeared in the "concrete" in the form of Dark Ages. Over the course of world history, these historical phases are rare. Between 3000 B.C. and A.D. 1000, there have been indications of only three such identified phases (2200 B.C.–1700 B.C.; 1200 B.C.–700 B.C.; and A.D. 400–A.D. 800/900) occurring in the world system from Northwestern India, East Asia, West Asia, the Mediterranean, and Europe (see Chew 2001, 2007, 2008). There is some evidence also of an earlier Dark Age around 3800/3400 B.C. but it has not been widely documented.

Some historians, such as Fernand Braudel (1972, 1981, 1982, 1984, 1989), have also echoed the aforementioned, that social, political, and economic factors in concert with the natural environment and climatological patterns are contributing factors giving rise to long-term and large-scale transformations or system crises. Several other historical scholars such as Desborough (1972), Snodgrass (1971), and Braudel (2001) have described and discussed qualitatively the conditions of life during these past Dark Ages highlighting the human-related dimensions such as economic, political, and social disorder. The trends are as follows:

1. Economic slowdowns and trade disruptions
2. Political breakdowns
3. Reduced social stratification and social simplification of lifestyles
4. Deurbanization
5. Increased human migrations
6. Population losses.

The panoply of social, political, and economic disruptions we see during these Dark Ages reveal the extent of the impact on the social and cultural fabric of human organizations that otherwise would not likely have taken place if times were good for human societies.

To briefly highlight on these qualitative transformations during these Dark Ages, besides economic downturns, political and trade disruptions, the reduction in social stratification, and the simplification of lifestyles are trends that do not usually follow past societal trajectories that historically follow a growth path. Whether these changes then determine a change in social organizational patterns is important to consider in our discussion of social evolution of human societies.

The same can also be raised for the increased human migrations we witness during Dark Ages or as we like to put it, social worlds in motion. The factors causing these migratory patterns are varied. Whether they are socioeconomic and political in origin or from climatological changes, these migratory patterns can lead to various systemic outcomes such as cultural upheavals, cultural assimilation, or even cultural homogenization that can then lead consequently to the transformations of the fabric of human societies.

With the increases in human migrations, deurbanization was the order of the day. Along with deurbanization, we also witnessed population reduction during this period of decline (Wolfram 1990). Because of climate changes, population losses were due not only to agricultural harvest failures leading to famines and starvation among the poor and the lower classes, but as well, to political unrest and wars that witnessed a high number of deaths and suffering. Disease pandemics also reduced population levels that in turn led to further deurbanization.

Beyond these socioeconomic and political transformations giving rise to reorganizations, the natural and climatological changes that usually appear during Dark Ages can also have a tremendous impact on the social structure of human societies. Fine weather during periods of growth prior to Dark Ages provided conducive conditions for heightened vegetation growth, and in turn, the rapid multiplication of herbivorous animals (gerbils), insects, and rodents because of the increased food supply. With the increase in the number of these organisms it led to the widening of the range of foraging for these pests and vermins. As a result, such a situation increased the likelihood of more contacts with humans. Rodents, such as rats, carry the plague bacterium that can easily spread to humans via fleas. As a result, we find the spread of communicable diseases and pandemics, especially during periods of Dark Ages (McNeill 1976; Keys 1999; McCormick 2001). Historically, this has been the case, in the various histories of the different regions of the world system. Take, for example, the devastation caused by plague pandemics during the Dark Ages of European history. The rise of pandemic diseases such as the plague during the Roman period, and especially during the Dark Ages from AD 400 onwards led to population losses due to the severe mortality rates that have been recorded (Millar 1981; Ruddiman 2003; Chew 2007).

DARK AGES AND SYSTEM TRANSITION

As we have stated in the previous pages, five thousand years ago the historical process of surplus accumulation occurred over wide geographic areas depending on the state of the existing infrastructure and technology. Timewise, in the very earliest part of world history we witnessed the emergence of human centers (urbanized communities) of surplus generation emerging in the valleys of the Euphrates, Indus, and the Hwang Ho (Frank and Gills 1992a; Chew 2001; Chase-Dunn and Hall 1997; Modelski and Thompson 1999). A historical process of this nature by no means expanded uninterrupted. As we have stated in previous pages, it was beset with crisis-like conditions and downturns leading to Dark Ages (Chew 2007). To surmount these Dark Ages historically, various directives were undertaken such as the intensification of technological innovations, political reconfigurations, intensified labor control and exploitation, or the incorporation of undeveloped geographic areas (hinterlands) whereby its bountiful natural resources and human communities were drawn into the circuits of the various centers of accumulation of a connected world economy.

These periods of Dark Ages, such as the one that appeared at the end of the Bronze Age, also produced qualitative changes emerging at the tail end of the crisis that provided the structural elements to enable the system to overcome the insurmountable crisis conditions. These qualitative structural changes, depending on which social, political, economic, and ecological dimensions that have generated the most pressure on systems reproduction, become the basis on which the directions of the socio-evolution of human societies are conditioned and shaped.

At this current stage of global human history, we are, as in the past, again experiencing such a phase of Dark Ages. This current global crisis—depicted by economic chaos, political unrests, global migrations, climate changes, pandemics and natural resource scarcities—is very similar in kind to that of the first two phases (2200 B.C.–1700 B.C.; 1200 B.C.–700 B.C.) at the end of the Bronze Age, and the third phase that occurred between AD400-800/900 commonly known as the European Dark Ages, though the latter crisis did not only occur in Europe but all over the Eurasian world system (Chew 2007, 2008). Interestingly, archaeologist Eric Cline in the preface to his book *1177 B.C.: The Year Civilization Collapse* draws a parallel between the collapse of the eastern Mediterranean at the end of the Bronze Age with our contemporary conditions.

Few will disagree that the world economy today is totally globalized in terms of economic, social, and political interconnections. Unlike the past, with the current globalized connections, a global economic, political, and ecological downturn has the danger of crippling the whole system, and will be a grave and major threat to the system's reproduction because the current global connections are now extensive space-wise, and intensive in terms of the global supply chains of manufacturing and food production. In this light, should it be a global economic crisis that the world system is experiencing, options for overcoming the global economic crisis will be addressed by the key core centers of the system, though such attempts are often hamstrung by political rivalry, competition, and regional conflicts. Efforts have been made, like in the past, to save the business exchange environment by enhancing, reconnecting, and rebuilding the trading networks via land and sea so that the global connections of trade exchanges can continue to expand uninterrupted, especially for the global hegemonic powers. Often, trade and military rivalries also follow.

Today, as in the past, the trade and military rivalries are pursued by the existing global power/s (Chew 2018). We witness such initiatives in the various competitive strategies to date executed by the various existing global powers. Timewise, today we can observe such actions on the part of the United States and the People's Republic of China. For the latter, its political rivalry and hegemonic ascent has even led political scientist, Fareed Zakaria (2020: 62), to amplify that "China's ascension to global power is the most significant new factor in the international system in centuries."

China's actions have been both global and regional in direction. Globally, its promotion and funding of its "One Belt One Road initiative," pursued for China's own hegemonic benefit is interesting from a world historical perspective. If we review the history of the world economy, it seems that the PRC is trying to re-establish the well-known land and maritime Silk Routes that linked the Eurasian world about 2,500 years ago that led to the emergence of the Eurasian world economy (Chew 2018). Historically, such linkages were disrupted over the course of world history as changes in global power hegemony shifted from one region of the world to another conditioned by the accumulative and political strategies pursued by the different global powers over the last five thousand years (Frank and Gills 1992a; Modelski and Thompson 1999). Given such a repetition of a historical trade structure, we can therefore expect political and military rivalries like the past to continue in the foreseeable future with political alliances forming across the globe to where the trade structure and routes traverse. For China, the expectation of capital accumulation and political domination are expected to be the order of the day. Witness its economic and political initiatives in Africa and parts of the Middle East.

If we drill down to some level of specificity, currently, at the regional level, China's ongoing initiatives in the South China Sea in terms of littoral occupation and territorial boundary claims have led to conflicts and challenges with its Southeast Asian neighbors of Vietnam, Malaysia, and the Philippines. At stake are the mineral resources and oil deposits located in this regional sea. Such Chinese territorial boundary claims do not acknowledge UN Law of the Seas agreement and serve to generate frictions and conflicts, and further reveal a flexing of the political and military muscles on the part of People's Republic of China.

As in the past, hegemonic power ascent has always been coupled with political and military rivalries of which at present are observed between the United States and China (Friedman and Chase-Dunn 2016). To counter China's strategic moves, the United States' actions in the Asian region as a global hegemonic power, has been to secure and strengthen its military and trade alliances with its Asian allies of South Korea, Japan, Australia, and lately India (Zakaria 2020). Militarily and politically, the United States has continued to assert its freedom of navigation for the United States Navy to sail through international waters, such as the Taiwan Straits and the South China Sea, and as well, the establishment of U.S. bases in Japan, South Korea, and Australia to counteract these attempted hegemonic efforts in Asia by the People's Republic of China.

DARK AGES AND STRUCTURAL TRANSFORMATIONS

In view of the current global crisis, are there possible changes in the structural elements that underlie the basis of the organization of our human societies that can provide the opportunity to transcend the global crisis that is restricting and endangering the reproduction of the social system and social order? Instead of turning our backs to the past and only casting our gaze to the future, perhaps we should consider the past and envisage what it can inform us. Following Cicero's trope, "history as the teacher of life," we can revisit the past to see what world history can teach us (Cicero 1942). If we return to the transition period from the Bronze Age to the Iron Age, we can delineate some structural developments that occurred during the late Bronze Age, especially in the eastern Mediterranean, which facilitated the socioeconomic and political transitions to a different organization of human society during a Dark Age. Briefly, the key structural elements (there are others as well) are the following:

a) The replacement of the base material from bronze to iron due to scarcity and political instability opened economic opportunities to locales that in the past had access to copper;
b) The utilization of iron in farming (such as the iron plough) led to increasing agricultural productivity, thus transforming social relations;
c) The shift in economic and political structures from command-centered palace economies to mercantile systems, especially in Europe and the eastern Mediterranean of the world economy impacted on the socioeconomic formations of this area of the world system, it led to political ("polis") development resulting in new political structures in the eastern Mediterranean further disrupting the political organization and the politics of the global system then (Chew 2007, 2008).

The emergence of the above structural elements at the end of the Bronze Age led to a system transition in the eastern Mediterranean that ultimately formed the structural features of the emergence of new principles of social organization (Chew 2008). What it meant then was a systemic transformation of the past ways of organizing the social formations in the eastern Mediterranean at the end of the Bronze Age when their continued existence was critically endangered.

The above structural elements are very historically specific for what happened at the end of the Bronze Age; but if history is our teacher of life, can we deconstruct the historical specificities depicting the Bronze Age transition, and translate them to more generalized characteristic patterns and parallel transformations to understand and analyze our current socioeconomic and political conditions? Let us proceed.

Briefly, for the *first structural change* whereby the base metal of bronze was replaced by iron during the end of the Bronze Age, one can deconstruct this shift to mean that because of Dark Age conditions with the limited availability of the base material (copper) being replaced by another; as a consequence it produced room for technological innovations and competition (Chew 2007: 100–105). Gordon Childe (1942) has commented that cheap iron with its wide availability, provided opportunities for agriculture, industry, and even warfare with the adoption of iron as the base metal replacing bronze. And with trade route disruptions and copper scarcity, the adoption of iron use spread further, for iron was available locally. It led to the development of local iron producing industries (Snodgrass 1971). The switch to iron as a base material gained widespread utilization throughout the eastern Mediterranean and Europe at the end of the Bronze Age. This provided the basis for the growth phase of the early Iron Age that followed.

Our current era is also witnessing a switch in the base material to silicon and carbon composites that, in the long run, will increasingly replace iron and steel. The new base material of silicon has permeated every aspect of our life, and with this Dark Age crisis whereby we are witnessing environmental degradation, global warming, disease pandemics, economic disruptions, political rivalries, global migrations, social and cultural strife; we envisage the coming of the end of the Iron Age on the horizon. The adoption of silicon as the replacement base material for iron and steel and its use leading to new production and social reproduction vistas is connected to the following second structural change.

Currently, what we are seeing in the production and reproduction of our life these days mirrors the *second structural change* that occurred toward the end of the Bronze Age when the switch of base material from bronze to iron led to increased productivity. The low cost of iron during the end of the Bronze Age because it was available locally facilitated its widespread use in agriculture and industry (Childe 1942; McNeill and McNeill 2003). Cultivation was made easier with iron plowshares in heavy clay soils that in the past wooden ploughs could not venture (Chew 2008). This enabled the rural communities to participate further in the economy beyond subsistence, and in maintaining a class of miners, smelters, and metal smiths fabricating the iron implements to reproduce material life. Such an explanation is also echoed by Heichelheim (1968) and Polanyi (1977), who have suggested that the widespread adoption of iron provided the opportunity for rural communities in south Russia, North Africa, Spain, Gaul, Germany, and Eurasia to work heavy soils with iron implements, thus increasing their production levels.

Another consequence of such a switch is that the urban elites in the Near East during the Bronze Age who in the past controlled the grain and other commodity trade suffered losses as a result of changing commodity prices, and also experienced similar setbacks with the falling demand of cooper and tin and the bronze trade which they also controlled. As a result, the social and political power structures were transformed with the formation of different regional centers in the periphery and in the Mediterranean. The opportunity for the farmers to farm in heavy clay soils utilizing cheap iron implements also provided the conditions for economic and system expansion following the end of the Bronze Dark Age where in the past these areas were not as productive, and some were ecologically degraded and overworked.

Stripped of the historical context, the adoption of iron at the end of the Bronze Age can be viewed as a transformative process with the replacement of the base material (copper) of an old technology that reached a point whereby economic growth stagnates as the existing technology and natural resource scarcity cannot generate further socioeconomic expansion. With the adoption of new base materials and socio-technological innovations, new vistas are opened for economic expansion and social organizations. It is similar for the current era, we can witness the vast economic expansions and technological innovations that silicon and carbon composites have given rise to for the current global economy, especially from the start of the computer age onwards to the use of social media of today. The adoption of digital technology has led to large-scale utilization of information that can be stored electronically in large data farms. With computation and application software designs including machine learning and artificial intelligence, these innovations made possible by the silicon chip have opened new horizons to the conduct of business, manufacturing, social and political lives that have never been conceived before. The use of carbon composites with tensile strength equivalent or higher than steel and aluminum means that the utilization of the amount of energy to power air transportation and other manufacturing processes will be greatly reduced. Manufactured objects that we utilize will be lighter to transport therefore requiring less energy and other natural resources.

At the end of the Bronze Age, with the adoption of a new base material such as iron leading to new technologies and increases in production, we also find the continued differentiation of commercial and economic structures from the political structures (Polanyi 1977). We begin to witness the third structural change coming into being. Instead of bureaucratic palace centered economic and political structures that dominated during the Bronze Age, we have the development of mercantile city-states where merchant enterprise replaced the palace-controlled exchange (Chew 2007). With this transformation, new forms of political powers and structures emerged. On the one hand, we have the emergence of a new political structure, the city-state (polis) in the Aegean. On the other, in the Near East, we see the continuation of empire type political structures where the rule was via direct political and military control. The rise of the polis in the eastern part of the Mediterranean was an attempt at transforming a kingship-type political arrangement to that of a community determined one. Of course, this did not occur in a conflict-free sociopolitical environment, it was beset with challenges, strife, and socioeconomic and political unrests, and especially via the rise of social movements.

Furthermore, it should be realized that the new political structure, the polis, as a social organization and political concept did not arise based on the basis of the adoption of a new base material and technological innovation alone. Other factors also precipitated its formation. The deurbanization and the loss of population in the urban areas during the Bronze Dark Age also engendered structural conditions for the development of the polis. In addition, with the scarcity of resources and the abundance of poverty leading to less hierarchical social structures, the groundwork for the development of the polis was also put into place (Chew 2007).

When we decontextualized historically the above historical events as to what happened at the end of the Bronze Age, the *third structural element* of the transformational process, suggests that the usage of the new base material and their associated technologies have led to different forms of socioeconomic-political relationships as the opportunities for different levels and forms of socioeconomic participation emerge. Returning to the current era, vast transformations have occurred in economic exchanges via the global World-Wide-Web, the use of social media, virtual and augmented reality, artificial intelligence, algorithms, machine learning, big data etc., that increasingly determine and structure our socioeconomic and political relations. Whether such changes will benefit some more than others needs to be measured, but it is clear that the socioeconomic and political networks formed have impacted every person's life at this stage of world transformation.

Besides, the transformations in socioeconomic relations due to the changes caused by the technological revolutions, other structural changes today such as the processes of deurbanization, migration, disease pandemics, and political unrests also emerge as consequences of the global crisis during this phase of system transition. Our current era witnesses the search for different socioeconomic and political arrangements as alternatives to the existing ones so that the crisis conditions can be surmounted. What results are alternative socioeconomic lifestyles and political arrangements veering away from the existing ones. In this way, we also find shifting cosmologies, ideologies, and so on, leading to a different socio-philosophical understanding of the world. In a socio-evolutionary context, we have the emergence of new structures and processes. Shifting ideologies resulting in changing life-practices, political participation, and political governance are attempted, developed, and even instituted as legitimate practices.

THE EMERGING NEW WORLD

Which of these above trends and tendencies have the potential to be impactful or already are? Within the scope of this book I wish to focus on some of the trends and ongoing structural changes that were covered briefly, and somewhat inadequately (in retrospect) in my book published over a decade ago, *Ecological Futures*. At that point in time, what we were observing in the virtual world, the development of artificial intelligence, and the spread of the various technologies associated with these new technological inventions were quite nascent and could not be fully anticipated in terms of the level and depth of transformations that have occurred. What is also surprising are the low level of social resistances and objections that surfaced then and continue even today. Following such trends, instead of resistance from progressives, what we witnessed these days is a further tight embrace by everyone, even from the potential contemporary "luddites" in light of the innovations and ease provided by the new technologies. For example, see Kirkpatrick Sale's (1995) *Rebels Against the Future* on this. Seduction and mesmerization by the various technologies developed have occurred, and is the order of the day, and in turn, these technologies seemed to be prevalent and encompassing of our lifestyles and choices. In some cases, these technical innovations such as the Internet, artificial intelligence, and virtual and augmented realities can provide opportunities to deal with the increasing scarcities of natural resources, disease pandemics, and the effects of climatological changes occurring during this period of global crises should not be automatically disregarded. They could possibly enable us to meet the critical contingencies that we will face as this phase of system transition proceeds. Recently, the widespread use of the Internet for meetings, education, medical health treatments, merchandise purchases, social discourse, and social exchange during the Covid-19 pandemic is a case in point.

How then do we live through this Dark Age, in other words, socially adapt and evolve so that socio-evolution can continue without changing drastically its natural course? What choices do we make? Considering the excessive global consumption, should we not question our current pretense that we are all trying to save our natural and social worlds? Our choices as well should not necessarily suggest that we become critical by demanding to save the world or the web of life by calling for an end to Capitalocene (see Moore 2016). Perhaps, in the interest of the human spirit during times of stress and that natural/social evolution continues, a certain level of pragmatism might be in order. We should not perhaps demand and reject in toto the many technologies and productive relations that have been generated in this "Capitalocene" phase's quest for value creation and accumulation. But instead, identify the various opportunities that these technologies have exhibited in the various areas that, I believe, will allow us to deal and surmount the Dark Age so that natural/socio-evolution will continue.

What is clear is that as the Dark Age deepens, the scarcity of natural resources required to reproduce our excessive materialistic lifestyles will increase, especially with the globalized conditions of the world economy these days, and the continuing growth in world population (Chew 2008; Ehrlich and Ehrlich 2004). In the past, population increases were tempered down by factors such as diseases, climate changes, civil unrests and wars, and it is increasingly hard to envisage that these factors that in the past had reduced population increases will be enough, especially with the many successes of modern medicine and health care (Schiedel 2017; Chew 2001, 2007, 2008; McNeill 1975; Ruddiman 2006).

To meet the scarcity of resources in light of exponential population growth, while at the same time continue the process of value creation so that social evolution continues, the structural changes that have emerged can provide a number of opportunities, if they are chosen and utilized effectively. This is couched on the premise that social needs are to be satisfied so that order and peace can be maintained. The structural shifts we are referring to involve the utilization of information and computerization which have provided the basis for further reproduction of the social system: the creation of a world of virtuality where value is created and consumed as commodities, and concomitantly, the process of dematerialization becomes a basis of organizing commodities produced and consumed increasingly in an organized networked social world (Mitchell 1999). For example, instead of producing and printing a book, warehousing it, and later selling it in a bookstore, the book is produced and published in electronic form, and stored in a server to be accessed or sold from anywhere in the world at any time. In sum, a net dematerialization. Furthermore, associated with this, is the process of miniaturization that will further reduce the use of material resources. Such transformation can thus address the issue of natural resource scarcities that will increasingly plague our socioeconomic and political needs.

I believe that these changes have the potential of alleviating some of the natural resource scarcity pressures, and even disease pandemics public health guidelines and protocols that are issued in the past in paper form can be digitalized and made more accessible. Of course, such shifts have to be calculated to evaluate whether a net gain in terms of natural resources can be attained or able to meet economic, social, and political necessities and demands. In fact, there are journals, such as the *Journal of Cleaner Production*, that are focused on publishing studies theorizing, analyzing and calculating the inputs and outputs of socioeconomic processes in relation to sustainable outcomes. The shift from iron to silicon and carbon composite base has generated the possibility of utilizing a lower proportion of raw materials and physical boundaries for producing commodities in view of the new technologies that are produced utilizing this new base material (Smil 2003; Chew 2008). The challenge will be to meet all the other human needs while utilizing fewer materials, physical spaces, and so forth. Commodity exchange is occurring in the virtual world whereby no material product is exchanged other than information generated via visual and auditory images, as exemplified these days in the virtual and augmented realities. The generation of such virtual worlds requires fewer material resources than would have been necessary if such activities occur in a common public physical space. Castells (2000) has called such socioeconomic and political interactions over cyberspace the culture of real virtuality. Symbols and signs produced and consumed in cyberspace form increasingly the whole cultural experience and generate both use-value and exchange value. As Baudrillard (1975, 1983) has so elegantly dissected, the logic of signification comprises of a functional logic of use-value, and an economic logic of exchange value besides the two other logics of symbolic exchange and sign value. Given such a construction, in the world of real virtuality, people's material and symbolic existence is captured and reproduced in cyberspace, and what is communicated in virtual images or appearances is not just communication of the experience, it has also become the experience (Castells 2000). It is substituting signs of reality for reality itself, and going from this, the generation of models of a reality without the origins of reality (Baudrillard 1983). Such simulations become the basis whereby socioeconomic, political, and leisure needs can be met where the provision of a tangible material product becomes unnecessary: the satiation of needs via images over cyberspace. No doubt, not all material needs, such as food consumption can be met through this, but a significant number of needs can be satiated. The move would be to shift toward an economy that entails little material products or physical structures—a "weightless" economy. We are at the beginning of a major system transformation: the Silicon Age will provide a vast array of possibilities whereby homes, workplaces, transportation systems, and electronic communications will be reorganized to produce new patterns, processes, and relationships in response to the contingencies we face in the social and natural environments.

With the beginnings of the establishment of the above new patterns and processes along with the technological products that have been introduced from smart phones to virtual and simulated reality devices, along with data storage in virtual space of cloud computing, we are beginning to witness the utilization of these technological infrastructures in numerous areas of socioeconomic and political exchanges. Here, we can witness the advances made in artificial intelligence for value creation via learning machines that have tremendous potential in lowering the level of inputs from energy to materials to satiate the consumptive needs of an ever increasing growing global population. In other words, these learning machines will provide responses and strategies for us to follow that will use less energy and other resources including a more efficient utilization of time-space. Such advances have also generated challenges for the political leaders in terms of increasing job losses, etc. that has been the outcomes for some time now. Routinized jobs have and will continue to disappear. Artificial Intelligence with the extensive utilization of algorithms will increasingly facilitate our total lifestyles and, as well, reduce the amount of time and resource utilization that are energy and resource consumptive because of the increased efficiency the learning machines are able to direct and provide.

Such above transformations do not mean all boats will be lifted at the same time, that is, not all regions of the world economy or countries will experience such changes equally. As in the past, there will be an unequal distribution of these productive technologies and infrastructures within and between regions and countries (Dauvergne 2020b). Countries or regions that utilize and invest in artificial intelligence and learning machines will be ahead, while others that do not will be left behind. Of course, with the current global inequality between countries not all the countries in the world system will be able to afford such investments even if they wish to. Perhaps with these new transformations, the discrepancy will be less between regions/countries in comparison to the past. That itself will be "progress" that has never happened in the past five thousand years of the world economy.

DEALING WITH DARK AGES' STRUCTURAL TURMOIL AND TRANSITION

Emergence of Alternate Consciousness and Ideologies

Making the choices of paths to follow in such turbulent times by no means is easy at the individual level or at the system level. If the decisions made at the end of the Bronze Age can be a guide for us to foresee what likely routes will be taken in the future during these times of crisis and transition, it will likely be multifarious. We expect that current existing dominant socioeconomic and political practices will remain in play, while new socioeconomic, political, and ecological relations and rationalizations will not have fully appeared nor have started to totally undermine the old ways. Socio-evolution is a slow process as five thousand years of world history have shown. The responses in the form of consciousness besides socioeconomic, political and ecological structural parameters that have emerged during this period of structural crisis appear mostly on the edges, and have not successfully challenge the past ways of rationalizing the existing ecological, socioeconomic and political practices.

Ecological consciousness is *as old as the hills*, it is not a stage reached when societal transformation has developed a rationality in the modern period that takes into account the environment via the passage of legislations according to the ecological modernization theorists (see, for example, Mol 2003). From world history we can witness that by 2700 B.C. from southern Mesopotamia through to India and China, there was an understanding of the dynamics of Nature and its relations with human communities. This ecological consciousness has been shown to exist in the philosophies and the historical epics of these civilizations (Chew 2001; Perlin 1989; Harrison 1992; Torrance 1998). The same holds true also for late Bronze Age Greece through to Greece of the Classical Period (Torrance 1998; Hughes 1974, 1994; Goldin 1997; Adams 1997; Westra 1997; Glacken 1967).

Times of uncertainty and distress such as the Dark Ages generate reassessment and reorientation. During previous Dark Ages, such as the last one that occurred during Antiquity (AD300-900), certain transformations emerged to address the socioeconomic and political disruptions and chaos then that in the long run were the determining factors in the development of Europe's social formation after the end of the Dark Ages of Antiquity in the tenth century. The emergence of Christian monasticism in the fourth century A.D. in Roman Egypt was one such transformation. Christian monasticism's appearance was a reaction to the vast excesses occurring then in all the socioeconomic spheres of Roman life (Chew 2007).

There are various interpretations and explanations for the beginning of Christian monasticism (Chew 2007). Some have viewed it with its beginnings in Egypt as a reaction to institutional Christianity and the imperial adoption of Christianity in the fourth century A.D.; while others have attributed its emergence to a call for a return to the values of Christian martyrdom (Workman 1913; Griggs 1990; Malone 1950; Frend 1972). Whatever the various reasons provided, the beginnings of monasticism can also be explained as an outcome of a structural transformation that started in the third century A.D. in terms of socioeconomic upheavals, political distress, and ecological crisis (Dodd 1965; Frend 1972). This period was also the start of the decline of the Roman Empire that dominated Europe and the Mediterranean. It was a time of widespread anxiety. Christian monasticism, from its spread throughout the Roman Empire and Western Europe from the fourth to the twelfth century, was one of the most continuous forces that played a determining factor of the most fundamental kind in the formation of medieval civilization according to some European medievalist scholars (Cantor 1994; Daniel-Rops 2001).

The monastic faith operating within a religious framework that requires retreat from inhabited urbanized communities of the times and seeking the return to one's original state, was a development that was incommensurate with the cultural and institutional religious practices of the Roman period. Urban life then in the Roman Empire was extremely natural resource intensive and environmentally degradative. Therefore, monasticism can be viewed as a movement offering an alternate worldview (*weltanschauung*) and a set of religious and social practices that was detached and religiously different from the ongoing practices of the times. Monasticism offered relief and salvation from the turbulent conditions of the Dark Age of Antiquity. As well, it can also be conceived as a social institution offering an alternate lifestyle choice for the reconfiguration of the social formation during the medieval period (Dunn 2000; Lawrence 1984). Especially during such turbulent and chaotic periods, Christian monasticism offered a framework that was quite different from existing conceptualization of socioeconomic and religious life and practices (Bergmann 1985; Athanasius 1980; Lawrence 1984; Goehring 1990; Daniel-Rops 2001).

Similarly, almost 1700 years later, in our current era of crisis that encompasses ecological, socioeconomic and political dimensions, the present tumultuous conditions have also sparked alternate worldviews (such as *Bioregionalism* and *Degrowth*) and lifestyle options. It is these reassessments and reactions to the socioeconomic excesses of our times that we will explore. Instead of running through the gamut of alternative lifestyle options that have appeared from the start of the twentieth century, this book will focus primarily on the specific reflections taking place in the production of our food, its distribution, and consumption, and the beginning reorientation of the socioeconomic view of growth via the concepts of *bioregionalism and degrowth.* I choose to focus on food consumption because it is one major sphere of our lives that cannot dematerialize totally as our species needs to consume food in a material context to reproduce our lives. The concept of *degrowth* as it pertains to our current food system is the need to reorient the manner and capacity we consume food materially and volume-wise. It is like what the late Bill Devall (1993) would suggest: *Live Richly in an Age of Limits*. This past call has been much lacking in the twenty-first century even by the environmental progressives. By no means is this ongoing rethinking conjuring up another new technologically oriented "scientistic" way of enhancing the food production system and methods of food consumption commonly introduced by the modernistic and postmodernistic practitioners to achieving progress and development. In fact, it is actually the opposite: it is a return to social practices (with an eye toward meeting the needs of a vastly more populated world) of the past that have been discarded and replaced when science-based industrial food production, distribution, and consumption were introduced to meet demands of urban living, corporate profits, and Third World development needs such as population increases throughout the twentieth century.

FOOD AS LIFE

Food and its consumption belong to one of the most important aspects of our social formation. Its impact not only is felt at the macro level whereby it determines a substantial part of the economy, policies of land use, socioeconomic relations, profits and incomes, and so on; at the micro level, it determines our lifestyle, our individual health, and our individual lives. Therefore, any structural change at the micro or macro level will have tremendous impact on the dynamics and operations of the social formation. In this regard, a changing worldview on food production, delivery and consumption practices in reaction to the increasing awareness of the various negative and degradative facets of industrial food production and its associated lifestyles can be transformative in the various structural changes and choices made in the political economy of food production, distribution, and consumption in the United States, for example. This changing worldview is not a call for a return to the past traditional form of farming practices that no longer can provide the needs of an overpopulated world. Its emergence underlines a shift in segments of our society toward a need to change our consumption habits and lifestyles following a growing awareness of the importance of considering the dynamics of Nature in our efforts to reproduce our lives, and in this case, our health and well-being.

In light of the above, the introduction of the concept of a 50- or 500-mile menu is an attempt to meet the challenge of sourcing food that are locally grown, thus promoting energy conservation, localism, self-sufficiency, wise land use, and so on. This emerging shift is counter to the current theme of globalization, technologization of life and hyper-consumption. It is an attempt to reduce the input costs (such as energy required to transport from the location where it is grown to the point of sale to the consumer) of food production. Coupled with this, is the theme from farm to fork with the daily farmer's market being held in urban centers as an alternative economic exchange locale for food purchases. Alternatively, with local restaurants entering into contract with local farmers whereby the farm produce is directly delivered to the restaurants is another way of following the 50–500-mile menu concept.

The farm share exchange practice of community-supported agriculture is also another example whereby consumers sign an agreement with a local farmer and entering into contract by agreeing to pay a monthly or seasonal charge for a weekly basket of fruits and vegetables grown by the farmer. In addition to fruits and vegetables, there are also farm share exchanges in meat and meat products. Such farm share exchanges also provide to the farmer a level of financial security that their produce will be exchanged thus reducing the farmer's risks that are always present in farming. Needless to say that all the food produced are organic and GMO free.

All these farming practices and structures can be enhanced by the artificial intelligence of learning machines that have the capabilities and capacities to generate learning algorithms that will streamline, enhance, and modulate the efficient flow of energy and other inputs, and provide the strategies and directions of the efficient and effective use of geographic space and time.

The above socioeconomic, political, and technological structural modifications can be coupled in some circles with increasing calls for the rethinking of growth and development. One such strategy is the proposal of the concept of *degrowth*. According to one of the *degrowth* network nodes based in Leipzig, Germany, *degrowth* as a concept covers a form of society and economy that aims to meet the well-being of all, and its purpose is to sustain the natural basis of life with a minimal impact on Nature. To achieve *degrowth*, it is argued that there needs to be a cultural reorientation. This line of thinking is very much in line with the concept of *bioregionalism* that was very popular in the 1970s–1990s among the environmentalists in the United States. How much of an impact such a *degrowt*h worldview will have remains to be seen? Clearly, we need to move beyond the concept of no growth, though a provocative idea by the *degrowth* proponents, is quite difficult, in my view, to reorient, especially with a global population that is expanding exponentially. The proliferation of the idea of *degrowth* will be dependent on how impactful the Dark Ages of collapse have on our socioeconomic and political systems. As the global collapse goes deep and long lasting, a demand for a reorientation in terms of a revised worldview will become more appealing, just like the rise of Christian monasticism during the Dark Age of Antiquity.

In what follows in this Dark Age of increasing natural resource scarcity and global socioeconomic and climatological crises, we will examine and contemplate the thematics and issues that I have outlined in this chapter in order that we can consider wisely the options available to us so that we can live through this Dark Age.

Chapter 2

Internet, Dematerialization, and Value

INTRODUCTION

We live in an informational world linked via global networks that encompass almost every part of Planet Earth. In the twenty-first century, this connected network infrastructure known commonly as the Internet is utilized increasingly by most of the world's populace, either voluntarily or involuntarily, to reproduce their daily lives. For the current period of system crisis and transition, the ubiquity of the use of the World-Wide-Web along with the technological advances made in computer hardware and software during the past decade present us (and our economic and political institutions) with the potential and opportunity (in terms of analysis and information) to deal with the global issues such as the growing scarcity of natural resources (minerals, energy, etc.), the consequences of global warming, the increasing deterioration of the landscape, disease pandemics, biodiversity elimination, and the availability of water to reproduce life.

In such an era of global crises, we must explore the panoply of options available to us to traverse the turbulent waters that we face in light of the conditions and outcomes that have emerged out of these ongoing global dangers. Notwithstanding the fact that there is unequal access and distribution of these resources to the world's populace, we cannot choose to avoid the deliberation on how the materiality of life can be sustained in light of the scarcity of natural resources to maintain an exponentially growing global population. Maintaining the ongoing system transition along its natural socio-evolutionary path to a pattern of socioeconomic and political framework that is sustainable without further generating intense global socioeconomic and political catastrophes will require a careful reexamination and analysis of the materiality of socioeconomic and political reproduction, and the reconceptualization of value in socioeconomic exchange relationships in view of present conditions. For the latter, to resolve current and future contingencies there is a need to abandon the historical socioeconomically established principles of the theory of value.

THE BACKBONE OF CONNECTIVITY

In the twenty-first century, the Internet as our basic communicative infrastructure enables us to exchange information between and within communities. This connectivity is made possible through the widespread use of silicon chips that are mostly housed in a computerized system such as a personal computer. Starting with the invention of the transistor in 1947, the integrated circuit in 1957, the planar process in 1959, and the microprocessor in 1971, these technical innovations have produced the modern-day computer and thus, ushering forth the Information Age. The invention of the microprocessor as the central processing unit of the computer allows the simultaneous storage of facts and information along with the instructions or programs on how to manipulate them. Furthermore, since the late twentieth century the microprocessor has also been installed in other technical apparatuses such as the cell phone, and other numerous devices that perform specific and varied functions to meet the socioeconomic and political reproduction of life.

In the early part of the twenty-first century we live in a connected world powered by microprocessors with digitalized information shared and distributed via wired/wireless communication lines. Global Internet protocol traffic in 2016 reached 1.3 zettabytes (Brynjolfsson and McAfee 2014). One zettabyte is a measure of storage capacity equivalent to 1,000,000,000,000,000,000,000 bytes of information. In terms of scale, the smallest unit is 1 bit of storage which contains a value of 0 or 1, and 8 bits of information make up a byte. The communication networks of this Internet infrastructure are independent of command and control centers insofar as information and messages can find their own routes along the network and get reconstructed at any point of the network. Such an open architecture allows any individual, firm, or agency to participate according to their functions and intentions. We have with this Internet infrastructure a system of interconnected information processing devices. Today, access to this World-Wide-Web is also no longer restricted to the microcomputer linked to a local area network like in the past, but is also open to other specialized devices such as the smart phone, smart refrigerators, personal assistants, and so on.

The computing power of this technological network is distributed in a communicative infrastructure supported by Web servers using common protocols of the Internet with access to other computers that serve as databases and application servers. Such a distributive system of information with worldwide access anywhere and anytime via a number of communicative devices means that it can be used for a wide variety of functions and needs from personal messaging and social networking, educating and scientific researching, policing, political activities, and war-making to e-business and manufacturing. In sum, every aspect of our daily lives, especially for those persons living in developed economies, is impacted by this usage. The ubiquity of such smart objects, available and used throughout the globe, presents opportunities (in terms of socioeconomic and political accesses) and dangers (loss of privacy) to the world's populace depending on the region of the world they reside, and their place on the socioeconomic and political ladder globally and within the national boundaries they inhabit.

This extensive panoply of utilization means that the information and data generated and exchanged are exponential and unimaginable, hence the proliferation of server farms and the increasing utilization of "cloud" computing to manage and store this information and data involved in almost every aspect of our lives. These days our lives are digitalized whether we like it or not. Furthermore, such development means that space and time can no longer pose a barrier to almost all our socioeconomic and political activities as compared to the our recent past, as evident by online purchasing from Amazon. Com to political campaigning via social media.

Given this cyberspace of global connections, the opportunity to use this connectivity to deal with global issues facing us, such as having a stable climate, wide availability of fresh water, diverse ecology, cheap and reliable energy, healthy landscapes, and a productive and healthy economy with high levels of employment, can now perhaps be addressed more efficiently and effectively. The connectivity plus the capacity to analyze, exchange, and share the information and data with the use of thinking machines employing artificial intelligence (AI) will offer us a range of alternatives to address, and perhaps, transcend the catastrophes we are going through, and what we will face further. It can provide us with the most reason–able choices and ways to live through this "moment" in our human history, and the socioeconomic and political reorganizations that follow.

In view of this highly digitized world of information, and with our current advances in analytical procedures and algorithms, we have the opportunity to address the global issues identified above that our forebears did not possess.

DEMATERIALIZATION: PATHWAYS AND PROCESSES

Over five decades ago, the Club of Rome Reports (Meadows et al. 1972, 2004; Mesarovic and Pestel 1974) projected of impending scarcities that the world will face if we do not control our population levels and our consumption patterns, especially for those of us located in the developed region of the globe. In a world of natural resource and material scarcities, the challenge will be to provide solutions that are feasible and reasonable to address these conditions, keeping in mind, that we do not want to use force to control the access to such resources. How do we proceed when these diminishing scarcities will continue to be needed for socioeconomic exchange, for after all we are material beings—we must "eat" to make history? The major challenge will be to meet always these growing human needs. Addressing this is crucial, especially when we continue to have an exponentially growing global population, an issue that is not being addressed seriously enough, in comparison to the energies and focusing that are these days devoted to the current discussion on global warming. Furthermore, it would be dangerous, or perhaps fruitless—especially with our history of consumerism couched within a market-based economy—to pursue the avenue to handle the natural resource scarcities issue by promoting or suggesting that one can live abundantly and richly during an era of limits by curtailing individual needs and wants through self-sacrifices and practicing "environmental correctness." In this manner, environmental correctness is applied to individuals who practice the recommended actions that will save our planet such as recycling and buying carbon credits for plane travel that have developed to become a general consensus and a normative obligation. Can these actors develop an ecological consciousness, so that they possess an ecocentric point of view? A parallel term to environmental correctness is political correctness whereby the actor acts according to the established rule or law, and not necessarily that the actor believes in the rule of law. Rather, it is followed as a normative obligation.

How then can we tackle this scarcity of natural resources so that socio-evolution can continue? The problem will be to meet the demands to reproduce life while utilizing fewer materials, less physical spaces, less energy usage, and so forth. The advances that have been made to date in the *miniaturization* of consumable commodities, and the innovations developed in the *dematerialization* process in the production of items and provide services required to reproduce our lives can provide opportunities for us to resolve the global scarcities issue (Mitchell 1999; Chew 2008; Smil 2016, 2014b; McAfee 2019). These two technologically innovative processes might assist us to continue to meet the varied human needs and consumption habits despite the decreasing levels of availability of scarce natural resources.

In general, real-world dematerialization refers to replacing "physical records or certificates with a paperless computerized system" (Smil 2014: 120). To be more precise and comprehensive, according to Smil (2016: 213), the concept of dematerialization is defined as the reduction of material used per finished product, or per unit of economic output, unit of power, performance, or service delivered. We can also consider dematerialization to cover the change in the amount of waste generated in producing a finished commodity. The concept of dematerialization is important, as the world's population is ever increasing, and the need to provide for its needs must be met. Therefore, there is less material required if we are to fulfill the needs. Simply put, the objective of dematerialization is to have a process of becoming free of physical substance ideally. In an age of digitalization, the replacement or turning *physical* objects such as books in our library and personal and financial records into digitalized files whereby each paper book or document is now turned into strings of ones and zeroes has already started. It is estimated that total global digitalized data in 2012 was 2.7 sextillion bytes or 2.7 zettabytes (Brynjolfsson and McAfee 2014). It was only half as much in 2011.

There are multiple paths to follow in the process of dematerialization (McAffee 2019). First, in the case of private business enterprises there should be attempts to find ways to use less of a given material in the production of a commodity for consumption. The utilization of less material input to achieve the same level of output in the production process. For example, the farmer using less land, water, and fertilizer to have bigger harvests or a manufacturer finding different ways to use less materials to produce a commodity. Second, to use a substitute substance over another. An example, which not everyone can agree is environmentally safe, is the replacement of coal using uranium to produce energy. A kilogram of uranium contains 2–3 million times as much energy as the same mass of coal or oil. Third, improve and optimize the efficiency of the production, circulation, and delivery of the commodity or services by improving the use of already existing equipment that the private enterprises already own. McAfee (2019: 111) cited how the commercial airline industry in America has improved its load factors through the efficient use of its aircrafts and other equipment. From 1971 to 2018, the U.S. commercial airlines have improved their load factors (passengers) from 56% to 81%. Finally, the manufacture of some commodities without the need to produce other objects. The obvious example these days of digitalization is the multiple functionalities of smart phones, such as the I Phone, that has in one product replace other products that we might have such as the camera, the dictaphone, the watch, radio, the yardstick, blood pressure monitor, thermometer, and so on.

The basic material for the reproduction of our consumptive needs has been based on iron, and especially from 1900 onwards, steel. Steel has no competition for its many exacting, durable, or heavy-duty uses (Smil 2016). A century later, this has changed with the availability of aluminum and titanium. The further utilization of steel, and in our case, iron as the base material is, of course, the costs of production of iron base products, the availability of the raw materials, the production costs in terms of energy usage, the amount of carbon dioxide emitted in the production process, especially with global warming, the need for lighter weight finished final products to meet the supply of other inputs to power the finished product, such as a car. In this regard, there is then the need to consider the substitution of steel and aluminum so that consumption can continue unabated keeping in mind the availability of the supply of natural resources, and the social normative rules and cultural perceptions of protecting the environment that might lead after all this consideration to find a replacement of the base material. The term substitution is usually seen as using an entirely different material to replace the existing widely used material, for example, in our case steel. Carbon fiber has the highest potential for weight reduction if cost of energy usage is a key criterion for the substitution, though it is six times the cost against the use of steel such as airplane construction like the Boeing 787 (Smil 2016: 206).

The invention of microprocessors and the personal computer has enabled the dematerialization process to occur not just in industrial production whereby miniaturization is also introduced to reduce inputs of material products in manufacturing the finished item, but as well also in everyday life of the world's populace. Books, services, billing statements, invoices, theater tickets, airplane boarding passes, timetables, and so on are digitalized and stored as information in databases and other storage devices. Submission of reports, examinations, building plans, product manufacturing blueprints, and so on are digitalized and exchanged through the Internet within a work setting and between work settings. Music, photographs, videos, art are now digitalized and exchanged, distributed, and consumed. What this means is the inputs toward a finished item that is exchanged have a relatively reduced amount of material inputs than before.

Besides the inputs of physical materials to reproduce our socioeconomic and political lives, the dematerialization process *can reduce the energy inputs* as well, to fulfill our needs, wants, and expectations. With the invention of virtual reality, E-art is an example whereby we can visit a virtual museum existing in cyberspace without the need to travel utilizing various energy sources to transport to the museum. E-tourism can also be considered as a possibility of fostering the dematerialization process. As it stands right now, with the growth of global tourism that is occurring nowadays, enormous amount of energy and materials are used up to support a global tourist industry, notwithstanding the damage done to the oceans, atmosphere, and landscapes. No doubt, with such proposals on virtual living there will be complaints of the lack of actual authenticity experienced through such ventures in comparison to what an actual physical visit will be for the individual. In response to such critiques, we should consider the *raison d'etre* of why we have museums. In layperson terms, it is to preserve past artifacts including the landscape for cultural posterity, and not for the current global population, or at least those in developed economies to consume it all too quickly to satisfy their self-indulgent fantasies. By transforming standard tourism into eco-tourism that is quite fashionable these days does not alleviate the problem either. All it does, perhaps, will be to clear some personal consciences, especially for those who have adopted the "environmental correctness" stance.

Reducing physical inputs and materials is a dematerializing process that will aid in meeting our natural resources scarcity contingency, there are other ways as well that will also reduce the volume and amount of physical inputs and energy streams. Via technical innovations whereby a new finished item's functional capabilities can replace the functional usage of several items that will engender the need to have only one finished product instead of several others. In other words, an innovative product such as a smart phone has displaced many other functional items we use in our everyday lives. The smart phone has replaced a number of other finished items we use in terms of functionality: communicative device, calculator, thermometer, address book, pedometer, flashlight, blood pressure gauge, video, and so on. By no means this tendency toward multi-functionality embedded in a single-manufactured commodity is the only solitary trend in the production process. For a market-competitive economy, there will always be the invention and introduction of new items that might redirect this tendency toward multi-functionality embedded in a singular product in other directions in response to lifestyle changes and market conditions.

Nevertheless, multiple functionalities in a single-finished product replacing a single-function item coupled with the miniaturization process can increase the lowering of the amount of inputs in terms of materials and energy in the production process, and thus assists in solving the natural resources scarcity contingency that we face. Going back to a smart phone, in 1990, a smart phone had an average mass of 600 g with about 11 million units in use globally, in comparison to a smart phone in 2011 weighing 118 g with nearly 6 billion phones in use (Smil 2016: 138).

Given the above, dematerialization and miniaturization processes that have started operating in our current era will continue to increase in terms of coverage. They have become part of the processes that reproduce our socio-economic and political lifestyles all over the globe, and especially in terms of overall penetration, they have advanced very rapidly in advanced developed economies. From what we can visualize to date, these innovations to reproduce the social and economic aspects of our world is similar to the adoption and innovations that occurred in the crisis and transition phase during the late Bronze Age. The adoption of a new base material, iron, in the late Bronze Age parallels the utilization of silicon and carbon composites in our era. With the wide-scale adoption of iron in the late Bronze Age, what resulted was the introduction of the iron plough that led to vast increases in agricultural productivity whereby in the past the wooden plough had limitations in heavy clay soil (Chew 2007, 2008). This shift in base material enabled the opening of vast acreage of farming lands in Europe, which in the past, were not available to farmers with wooden ploughs. Like-wise in our era, the emergence of the informational society has generated the potential and possibility of boundless interactions through cyberspace in our socioeconomic lives, expanding and enhancing the production process, and overall output of the societies and civilizations (Castells 1996, 1997, 1998; Chew 2007, 2008). The dominance of information technology companies in terms of the market value of their stocks listed on all the global stock exchanges from New York to Tokyo needs to be underlined. In 2017, U.S.-based tech companies registered the highest market capitalization (Robinson 2018: 6). Companies like Apple with a market capitalization of $730 billion, Alphabet (Google) with $581 billion, and Microsoft with $497 billion, to name a few, are some of the most valuable companies in the world. The power that these companies have and the owners and personnel that manage these corporations have increasingly become part of the power elites in America that C. Wright Mills (1956) wrote about in the twentieth century.

The boundless opportunities of the cyberspace will not mean necessarily the utilization of less resources to satisfy needs and wants within a framework of scarcities. As the established market economy principle, the TINA (There is no Alternative other than Capitalism) syndrome, has not been replaced by another alternative to determine economic relations, endless commodity production that has been dematerialized and miniaturized to sustain surplus value accumulation will still put extreme stress on the available natural resources that are required for production. In this light, re-conceptualizing the goal and aims of the human endeavor, and our place in this ecological world can perhaps offer some new directions.

VALUATION: THE LAW OF VALUE OF SYMBOLS

Since the dawn of human civilizations, the exchange of commodities (trade) between individuals, groups, cities, kingdoms, and nation-states has been an ongoing process whenever there is surplus production. This economic exchange has, to a large extent, involved commodities of a material nature; of course, the materiality of the commodity is dependent on the particular phase of the circulation of what is known as "capital" accumulation. To put it in another way, depending on the phase of capital circulation, the materiality of the commodity only appears when raw materials and natural resources are introduced in the production process along with the utilization of energy to manufacture the material product. This material product, is, then exchanged either for another material commodity or for money that can be in various forms (gold, silver, copper, paper, etc.) mostly depending on the time and space of civilizational development (Marx 1959, 1972; Harvey 2018). The underlying factor of this exchange process is value, which is the expression or representation of the social relation of the exchange. Money as value, in whatever form, is the material representation and expression of the relation of economic exchange process. In this sense, what is often discussed in terms of the valorization of capital to describe an advanced developed economy is the expression of value-in-motion. This value-in-motion depicts the process of how money-capital is transformed into commodity-capital through the labor process, and then through the exchange and circulation processes back into money-capital for further production and investments. Various groups of players are involved in this circulation of capital, and thus the exchange process is complex, especially in advanced developed economies.

The determination of value in a commodity is dependent on its usefulness, the various inputs of natural resources and energy involved in its production, the labor and transportation costs, and the numerous infrastructures (roads, airports, ports, etc.) involved in bringing the commodity to the market for exchange, along with the costs of enforcement and regulation activities, and so on, that the States provide to ensure a framework of lawful and legal exchange.

Therefore, in light of these numerous activities, for the advanced economies as we have stated above, a multiplicity of groups ranging from owners and managers through to workers and state employees (bureaucrats, police, and the military), are actively participating in this valorization of capital to ensure that these circulation processes are not interrupted. Structural infrastructures and legal frameworks circumscribed this valorization process to guarantee its legality and its cultural significance. The reproduction of cultural lifestyles and socioeconomic well-being are, thus, ensured. All these various moments of value-in-motion occur within a sociocultural context of mores and norms of the civilizations and nation-states. The socioeconomic and cultural expectations are that the process of exchange is to fulfill individual needs and wants in the reproduction of our lifestyles. In other words, to reproduce the required use values for the continuation of everyone's lives. In an era of scarcities of natural resources, water, arable land, and the exponential increase in global population, not all needs and wants can be distributed equally. Like in the long-term past, only the top tiers of the population of the globe will have the capacity to reproduce the use values of commodities for the reproduction of their lifestyles. Even for those in these top tiers, there might not be enough resources for them as well. Therefore, cultural values and meanings of socioeconomic practices will need to be adjusted, if not changed broadly, so that the reproduction of socioeconomic life as expected can continue. To survive the current global crisis, we need to reorient our mental conceptions of materialistic exchanges.

All of us, especially those living in advanced developed economies, have understood and accepted that the reproduction of our socioeconomic and political lives involved exchanges processes that mostly involved material commodities. There are, however, other aspects of our social reproduction that involve exchange of services that are not material based in terms of what is exchanged other than money. Here, I am referring to an exchange whereby one receives a service without getting a material commodity. From a cognitive conception point of view, if we need a service, and we exchange for that service with money, we have agreed to an exchange without questioning or wondering the legitimacy of exchange value of this type of exchanges, i.e., service provision. Then again, should the type of exchange be made that is not for a service but for a material commodity such as a paper book, we would expect that the exchange would involve money for a paper book. Though with the major leaps in computer technology that have taken place, such as in the digitalization of material products and the storing of these digitalized files in electronic storage facilities in server farms, the exchange process will still involve money but the purchaser, in this case, has agreed to accept an E-version of the paper book, instead of a paper book, as an equivalent in terms of exchange value. The seller of the E-version of the book realizing that the purchaser will not be receiving a paper book has priced the book at a much lower cost, and at the same time, this has the effect of also prodding the buyer to accept the dematerialization process that is starting to pervade aspects of our lives in which material products formed the reproduction of our lifestyles.

Increasingly, the virtual world or cyberspace is where a large proportion of our socioeconomic and political activities will take place. Natural resource scarcities, the effects of global warming, the devastation of natural landscapes, and the spatial crowding and urbanization will force us to move more and more into cyberspace in order for us to continue to reproduce our lives as we know it, and perhaps, not as we like it. We will increasingly accept "weightless" (dematerialized) products, if we have not done so already, in our economic exchanges. Given such circumstances, in this cyberspace world, people's material and symbolic existence are captured and reproduced. In our exchange processes with other individuals or with financial and business entities, and also with political institutions, we have or will accept(ed) this virtuality. This, perhaps, has or will transform(ed) our notion of value, which has in the past been anchored in a materialistic reality.

Given this, we can see that technology, including scientific knowledge, has an important impact on the overall value generation in the circuit of capital in motion. Both the hardware and software that have been created have impacted on every aspect of our lives these days. One of the most significant items to multiply exponentially is the appearance of what is known as an "app" (application software) that forms the basis of our socioeconomic and political interaction in cyberspace. For example, in 2019, the Apple App Store alone has about 2.2 million apps for downloading to be used in our smart electronic devices. These application softwares have determined almost every aspect of our lives in almost every environment that we live in to reproduce our lives. From a value-in-motion circuit, there are different "apps" that have been developed with the premise of increasing efficiency, economy, and effectiveness in each phase whereby value is generated and exchanged. No doubt, most of us do not visualize it as such, other than seeing these "apps" as enhancing our quality of life, and especially increasing our efficiency, economy, and effectiveness of our social reproduction. The ease these "apps" have slipped into the way we live have opened up vistas on how we can organize our socioeconomic and political lifestyle organization. Developed with different premises in mind, these "apps" have eased, in most cases, our everyday routines. From social media apps that inform, direct, enable us to exercise our needs, wants, and aims in our personal lives to apps that allow us to conduct economic exchanges for material and non-material commodities that reproduce our daily lives, the widespread permeation of these application software have captured us. Other than some die-hard technological Luddites, there are very few persons these days that do not use these apps to live and reproduce their lives.

In addition to these application softwares, the development of what is termed machine learning will also add value to overall circuit of "capital" in motion. It is part of an overarching science known as AI that is concerned with intelligent algorithms, with machine learning being one of its subfields that are devoted to algorithms that learn from data. The science of AI consists of algorithms that have the capacity to provide choices to solve problems that we face (Domingos 2015; Barrat 2015; Tegmark 2017; Husain 2017; Ford 2018). Some have stated it is the next human invention that is supposed to transform our lives, and with this, it will impact on the value of materialistic and non-materialistic commodities that we consume from food to entertainment. It will help us to make choices or present us with recommendations so that we can make our decisions based on vast amount of data collected globally on each specific issue that we will face from global warming to the cure for diseases. No doubt, not all the nations of the world will benefit from this development. As always, those nations with developed advanced economies that have invested in AI research will be able to maximize the socioeconomic and political transformations and development. Translating downwards from the nation-state level, those companies and institutions that have devoted resources to AI will also enhance the value of their activities in the overall circuit of the valorization of capital.

If AI is projected to continue to have more and more of an impact in our daily lives from the social sphere through to the political sphere on a global scale, the valuation process in determining the socioeconomic exchange will increasingly be located in the technological knowledge realm. Given this, it would be safe to say that advanced training within a multidisciplinary context becomes the determining factor in the distribution of value. How just and equal that distribution becomes is circumscribed by the politics of a given social formation. To date, for over five thousand years of world history the distribution of value has often favored those at the top of the vertical ladder.

Chapter 3

Artificial Intelligence, Learning Machines, and Virtuality

INTRODUCTION

During an era of abundance and surplus, it is not difficult to select choices of suitable paths to follow for the overall reproduction of life at the level of the individual or the group. Unfortunately, we are not at this bountiful stage of the planet's reproduction. We are in an environment of scarcity and stress. How then do we make appropriate intelligent choices so that we all can traverse this system transition without having to endure the foreboding consequences that so many of the experts and politicians have proclaimed and projected for us? Should we still consider reasoning within a modern mindset of hope in the belief of human rationality and progress, or should we dwell in the worldview (*lebenswelt*) of postmodernity? Postmodernity is the phase that follows modernity (Harvey 1990; Habermas 1981). It is a phase that does not subscribe to the perspective of a universal truth. Rather, the phase celebrates a *mélange* of viewpoints that do not purport to have a singular basis of rationality or truth with a capital "T." In this regard, diversity is celebrated (Habermas 1987; Harvey 1990). Alternatively, could we be pragmatic by adopting both worldviews (*lebenswelt*), and pick paths that would seem to be most likely for us to transition successfully in light of the available historical information? We do have at least one past transition that we are aware of, that of transitioning from the Bronze Age to the Iron Age, which we can consider, and perhaps, to guide us as to what the challenges the current transition will be like.

Can History be the Teacher of Life like what Cicero (1942) has said so many years ago? If History can be our Teacher of Life, we must also realize that past historical actions have shown that sticking steadfastly to a singular worldview (*lebenswelt*) have resulted in social systems that are not that enduring for long. The more recent example of the demise of the former Soviet Union is a case in point. There are, I am sure, many other examples that one can point to as well. Given this, if we operate with a modernity and postmodernity mindset, it would be very beneficial, therefore, for us to make choices that have considered a wide array of possibilities based on the analysis of a wide array of information and options produced by intelligent algorithms from machine learning. That is one of the paths that I want to propose in this chapter for living through the Dark Age.

The main task ahead of us in this system transition is to make decisions that will enable and *ensure that social and natural evolution continue unabated.* At a moment in world history that has been categorized as an age of information, we need to harness the technology and scientific knowledge that have been generated to inform us as to the most reason–able choices to make to ensure a transition to a social system of socioeconomic and political organization that will safeguard our harmonious relations with Nature; even though in the past five thousand years the latter relationship has been more exploitative than harmonious (Chew 2001). In what follows, I propose to highlight some of these technologies and scientific knowledge, and how they will assist us in this environment of growing scarcity.

ALGORITHMS

According to an Artificial Intelligence (AI) expert, Pedro Domingos (2015), our lives are already suffused with many features of machine learning, which is a subfield of artificial intelligence "devoted to algorithms that learn from data" (Husain 2017: 21). The term itself, "artificial intelligence," was popularized at a conference held at Dartmouth College in 1956 that hosted researchers on a range of thematic areas from language simulation to learning machines (Manyika et al. 2018).

As noted by Domingos (2015: xii), as soon as we wake up daily to our alarm clock, different aspects of machine learning are already impacting our lives:

"Your clock radio goes off at 7:00 a.m. It's playing a song you haven't heard before, but you really like it. Courtesy of Pandora, it's been learning your tastes in music, like your own personal radio jock. Perhaps the song itself was produced with machine learning. You eat breakfast and read the morning paper. It came off the printing press a few hours earlier, the printing process carefully adjusted to avoid streaking using a learning algorithm. The temperature in your house is just right, and your electricity bill is noticeably down, since you installed a Nest learning thermostat.

As you drive to work, your car continually adjusts fuel injection and exhaust recirculation to get the best gas mileage. You use INRIX, a traffic prediction system, to shorten your rush-hour commute, not to mention lowering your stress level.

At lunchtime you walk down the street, smart phone in hand, looking for a place to eat. Yelp's learning system helps you find it. Your phone can even anticipate what you're going to do next and advise you accordingly. For example, as you're finishing lunch, it discreetly alerts you that your afternoon meeting with an out-of-town visitor will have to start late because her flight has been delayed."

This brief short excerpt by Domingos (2015) shows clearly the extent machine learning has been involved in our daily routines. For our purpose, we are not going to underscore the convenience and ease machine learning has assisted us in living our daily lives, which, I am sure, a lot of people, especially those living a busy and hectic life, appreciate. What needs to be stressed with regards to the above excerpt is the *amount of energy and resources* that should have been conserved as we individually utilize machine learning to assist us in choosing efficient and economical options in going through our daily routines. It is how AI can assist us in dealing with *the global and local environmental contingencies* we face, presently and in the future, instead of just viewing AI as a technological development to better our socioeconomic lives by providing different functionalities to ease our daily living, and generating new commodities for private businesses and enterprises to grow exponentially their profits and dominance in the global marketplace. Of course, the amount of savings in energy and natural resources need to be carefully calculated, though according to Smil (2014), such tasks are often complicated and difficult to measure and evaluate. Environmentalists need to address in this era of scarcity of resources such as energy, and so on, how the availability of timesaving, economical, and effective options through machine learning will allow us to transition more smoothly during a phase of scarcity. It is important to note that the level of savings in energy use that can be achieved, if these options are made and utilized at the national and global levels by policy decision makers, in the form of laws and regulations will deal seriously with the issue of global scarcity that we all face. If not, the scarcity issue will continue to amble along leaving "the long emergency"—a term derived from James Kuntsler's (2005) book—that we have been facing to the vagaries of market forces, and depending on the state of transformation of the individual psyche to follow through with environmental consciousness or environmental correctness. However, if options are chosen wisely based on information and analyses generated from AI, it will mean a big step forward for all of us to experience a more smoother system transition.

To reproduce life in the overall sense requires *energy*. According to Stanford University archaeologist Ian Morris (2013), the foundation of social development is energy capture. By energy capture, Morris (2013: 53) meant three broad areas that are necessary to reproduce life, and the measure of them determines the state of social development:

> *food* (whether consumed directly, given to animals that provide labor, or given to animals that are subsequently eaten);
>
> *fuel* (whether for cooking, heating, cooling, firing kilns and furnaces, or powering machines, and including wind and water power as well as wood, coal, oil, gas, and nuclear power);
>
> *raw materials* (whether for construction, metalwork, pot making, clothing, or any other purpose).

These three categories cover the major portion of the gamut of global scarcity issues that we face in this era of crisis and transition. Making the right choices as to their extraction, processing, and consumption is crucial, and the way to make the appropriate decisions needs to be done in relation to information and options available to us. It means at this conjuncture, the shift to AI machine learning. Because of the scarcity that is expected, especially in terms of energy for fuel to reproduce our varied lifestyles, it would be prudent to focus on how machine learning can provide us with information, options, and possibilities, so that our usage of fuel to meet our needs will enable a smoother transition than we would otherwise (Smil 2008; Friedrichs 2013; Brown 2009, 2015).

How then do we stretch out the existing fossil fuel supplies, and at the same time, transition more to alternate energy sources which is the favorite mantra these days of mainstream environmentalists, in view of the environmentalists' overall preoccupation to address global warming as if this is the only danger that we all face. Expending energy use of fossil fuel in the immediate term should be addressed more forcefully. In a world of highly urbanized living whereby 50% of the world's population now live, driving to work, especially in advanced developed economies, continues to be the norm. This mode of transportation increasingly utilizes the GPS technological software to assist commuters in the Unites States to travel from home to work. The traditional version of the app has been able to map all the different routes comprising of different highways, surface roads, and so on from point A to point B. This early traditional version, however, does not tell you in real-time the location of traffic jams, accidents, road closures, and other factors that affect travel time during the time of your commute. When asked to calculate the travel time, this traditional version just maps out the most direct route using highways, since they are the ones that have the highest speed limits and informs you the time it will take. No doubt, this version of the GPS app will reduce your travel time, but it does not provide real-time data of traffic conditions. A more useful GPS app can know where the other cars are, and how fast are they moving on the same route. This is done via smart phones or GPS app built in the cars that can act as sensors uploading constantly to cloud servers their location and speed. With this, the GPS app can provide real-time on road and traffic conditions to the driver with constant provision of option/s for re-routing so that travel time on the route will satisfy the time that was provided initially to the driver at the start of the journey. To do this, the cloud servers are utilizing the digitalized GPS maps, the constant updates of traffic conditions, and a set of sophisticated algorithms to generate the options of driving routes. The latter shows how AI can not only lessen the time traveling by avoiding traffic jams, and therefore, decreasing the overall utilization of energy for traveling between point A and point B. Clearly, widespread utilization of this type of GPS app will assist in overall reduction of energy use.

The above example offers us a glimpse of how substantial savings of energy utilization and time for an individual consumer to travel from point A to point B when the GPS app is used. With the ability of machine learning being able to scan vast amount of data and come up with options and possible solutions to reduce energy consumption in the use of transportation to reproduce our daily living; there have been efforts to reduce overall energy use further besides the loss of time in terms of work productivity due to commuting.

With the collection of data, in some ongoing studies, that covers the duration of traffic volume over the course of a year on each specific day of the week and time of day for an urban environment will allow traffic planners to coordinate the timing of traffic lights. This will ensure that there are no traffic snarls that commuters must wait needlessly long at traffic lights even though the traffic is light. I am sure we all have experienced this. With digitalization, the data collected (big data) can be read easily, thus increasing the capacity of machine learning. The deep learning methods of machine learning that are ongoing will advance further in view of such a wide array and volume of data available. For example, Volkswagen AG in collaboration with Nvidia Corporation of California are developing deep learning algorithms to analyze mobility patterns of traffic in several urban environments so as to optimize traffic flows in order to minimize time taken for travel to reach your destination (Castellanos 2017). The information collected from traffic lights, pedestrian crossings, and road construction is analyzed through deep learning to discover new patterns and optimum traffic routes. A test conducted in 2018, came up with the traffic routes for thousands of taxis within fractions of a second to optimize traffic when the test used quantum high-speed computers that normally would take a normal computer three-quarters of an hour to complete the same task. The availability of cloud storage and its easily accessibility will further increase the reliability of the analyses and options produced. Given this complex level of digital infrastructure available, overall energy savings will further increase when autonomous automobiles are introduced as a mode of transportation for commuters or any person requiring any carriage from point A to point B. The functioning of autonomous driver-free vehicles relies extensively on AI for its operations.

Clearly, the above development in assessing traffic flows, traffic light coordination, and so on will be even more important if there is a switch to commuting to work using a personal vehicle instead of the use of public transportation, especially, when there is a disease pandemic in place. We can see such shifts in personal use of vehicular transportation for the reproduction of social life during such pandemic conditions as we witnessed during the Covid-19 pandemic of 2020 whereby we notice the change in the type of vehicular transportation used (personal cars), especially in the developed countries such as the United States.

Overall, AI has three types of capabilities that can assist us in managing the transition in light of the levels of scarcity that we are facing and will face (Davenport 2018; Ford 2015). The capabilities might not all be as useful to us to deal with the environmental issues that we will be confronted. The first type of capability that AI can provide is the automation of production processes through robotics or robotic type automation. This by far has been implemented in business operations such as replacing credit cards without human intervention to reading legal documents to extract contract provisions. The second capability is one which uses algorithms to detect patterns in big data and interprets its meaning. This second function that covers machine learning provides cognitive insights for the user. The third function covers engagement with the client. In this context, in business activities, AI is introduced to respond and interact with clients in the natural language of the client. Under this capability, for example, health treatment recommendations can be proffered in an interaction with a customized care plan that considers individual health status and previous treatments.

AI has been adopted for business operations, inventory surveillance, voice and facial recognition, and so on for the marketplace. It is still, however, in the early stages of its deployment by the private business sector, and as well, the public sector. Chui et al. (2018) in a McKinsey and Company discussion paper, has revealed that in a study of 400 use cases in the private sector across nineteen industries and nine business functions; their analyses project that AI has the potential assisting in the operational functions of these industries, especially in retail, transport and logistics, and travel. Mostly these are in marketing and sales, and operational functions like supply-chain management and manufacturing.

In the above cases, AI is used to solve classification, estimation, and clustering problems in a market exchange economy. These three analytical techniques together can potentially enable the creation in the global economy of between $3.5 trillion and $5.8 trillion in value annually. According to Chiu et al. (2018: 17), per industry, AI's potential value amounts to 1%–9% of its 2016 revenue.

An example of how AI can impact on the travel industry, and the industry's potential, is the following: Hawaii's Tourism Authority working with a major online travel company using facial recognition to monitor travelers' expressions through their computer webcams thus enabling it to deliver personalized offers to improve revenue related marketing and sales (Chiu 2018: 17).

AI has also been utilized by a utility company, Pacific Gas and Electric, in California to assist its operations in reducing the risk of wildfires. The strategy is to monitor distribution poles and transmission towers in terms of possible equipment malfunction that result in fallen power lines or sparks that can generate wildfires. This monitoring process is comprised of drones that are deployed to fly over power lines, and with computer vision from the drone and machine learning to spot potential trouble spots and equipment issues. The computer vision can be trained to classify what something is supposed to be and spot if there is something broken. Machine learning can undertake the analysis to see there are signs of changes or corrosion. In this way, it allows PG&E's operational staff to decide the work repair schedule, and perhaps even to shut off the power if severe weather conditions persist. In this way with the computerized drones and AI the expense of having to utilize human resources and energy for monitoring is made more efficient. Access for repairs is also made easier, especially in difficult terrain, where it is difficult for maintenance crews to reach, and especially in severe weather conditions (McCormick 2020).

Clearly, the private sector's business activities and functions have adopted AI in their operations. Wider adoptions across industry and sectors will certainly help to reduce the demand for natural resources and energy that are crucial for our global socioeconomic development. AI's varied capabilities and capacities can conduct various aspects of a private business enterprise's operations, such as monitoring preventive maintenance as well as optimizing its production and assembly processes. This basic ability of AI's functionality can be applied across the board of the private enterprises' business operations. All such monitoring and analyses can minimize spending on utilities and raw materials and can lower production costs. For example, Chiu et al. (2018) have estimated that it can reduce overall production costs by 5%–10% of the 400 cases of the 19 industries they have studied. In view of this, the private sector of the advanced developed economies should consider implementing further machine learning while saving operational and production costs, this sector can also help to reduce the stress on the scarcity of natural and energy resources that we all face currently and in the future. What has been discussed above is an exemplar on how from an environmentalist viewpoint, AI can be employed to solve the scarcity of natural resources and energy, and perhaps allow us to transition less chaotically during this period of upheaval. The potential of AI goes beyond providing solutions to solve energy scarcity issues, it can and has been utilized to addressing quality of life issues in the area of health, diseases, food production, and wars.

THE DIGITALIZED WORLD OF VIRTUAL EXPERIENCES

In the twenty-first century, economic exchanges involve not only commodities of a material nature, but as well digitalized products via visual and auditory images characterized by virtual and augmented realities. In value terms, symbols and signs become the significations in a world of real virtuality. Our material and symbolic existence are captured and reproduced in the cyber-world. And according to Castells (1997, 1998, 2000), what is communicated in virtual images or appearances is not just the communication of the experience; it has also become our experience. In a Baudrillardian (1983) fashion, by substituting the signs of reality for reality itself, the generation of many models of reality leads to a "reality" minus the origins of reality. Such digitalized simulations generate in our mind the provision of a "tangible material" product for the satiation of our needs in mainly an economic exchange. Therefore, in this case, the theory of value, as we have elaborated in the previous chapter, covers not only products that of a material nature, it also includes the exchanges that comprise those of signs and symbols of the cyber-world. This transformation conduces us to accept in a market exchange economy the exchange of commodities that are not composed of material matter. Such a shift in the process will alleviate the scarcity of natural resources and the need for dematerialization that we all face during this global crisis phase, and the system transition that we will need to navigate.

The virtual world now focuses on the development of two types of reality experiences. Virtual reality (VR) immerses the user/person in a totally digitalized environment, whereas augmented reality (AR) overlays virtual objects onto a real-world environment. With the arrival of AR, digitally constructed artifacts can be superimposed on real images thus bringing objective reality and perceived reality into even a sharper focus. Like the VR products that have been introduced, AR products can give more flexibility for various transactions in the social, economic and political spheres.

The VR and AR products that have been conceived and made available for our "consumption" have begun to appear and used in the various spheres of our lives, and many more are in the research and testing pipelines of established and start-up Silicon Valley companies such as Facebook/Oculus, HTC,

Microsoft, Apple, Google, Magic Leap, HQ Software, VIRONIT, NEXT/NOW, VR Vision, Corporate Pop, and Program Ace. The vast potential of what virtual and augmented realities products can alleviate our scarcity problems by providing functionalities and coverage that can satisfy our needs and wants in the social, economic, political, health, and sexual spheres, is yet to be realized (Rubin 2018; Bailenson 2018). This sphere of virtual and augmented realities products that are reasonably priced for overall public consumption has yet to be fully rolled out. Initial reports suggest their use individually for personal lifestyle consumption is going to be exponential, especially in terms of the various functionalities these VR products can offer (Bailenson 2018; Scoble and Israel 2017; Rubin 2018). The insertions of virtual or AR in products we use have started to roll. However, our interest here is primarily focused on the array of offerings that, in our view, will help to address the scarcity of resources to reproduce our lives in this ongoing system transition.

Virtual reality, according to Stanford University VR researcher, Jeremy Bailenson (2018: 5), has the potential of making us "better people, more empathetic, more aware of the fragility of the environment, and more productive at work." It gives us the sensation of "being there" wherever the VR program transports you to. Your interaction with the VR program is very similar to how your perceptual systems interact with the real physical world. Hence, the "experience" would seem to be as identical to what you normally experience (the presence) with the physical reality (Rubin 2018). It is, therefore, not really a media experience but understood like an actual experience to your mind (Bailenson 2018: 46). When in a VR environment, what we see is not framed by a TV or movie screen whereby we see perhaps 180 degrees what the director or editor of the video directs us to see. Instead, we see the whole virtual world as we are immersed into the total digitalized environment.

As a tool, it can be utilized to improve the quality and reach of education, open up the world for travel especially for those who do not have the financial resources to do this, and to be able to share our experiences with others who are not living in the same geographic locale as us. There are countless numbers of ways this technology can be applied to all spheres of our world (Rubin 2018). Almost every aspect of these functionalities that VR technology can provide will mean the *opportunity* to use fewer natural resources and energy to reproduce our needs, wants and expectations, especially in the economically developed countries and its populace.

When put into play, VR gives us the illusion of "presence" as it can transport us to anywhere geographically and organically. The latter meaning that we can slip into any living organism and experience their worlds! This experience is very much different than watching a film or playing a video game. In playing a video game, for example, we can differentiate what is playing on the monitor and reality. But with VR the gap between real experience and mediated experience is very much smaller (Bailenson 2018: 6; Rubin 2018; Scoble and Israel 2017). You are immersed in the digitalized world that has been programmed and you gained access to experiences that you cannot usually have. For example, according to VR researchers you can slip into fishes and swim in the coral reefs, see the world from the perspective of a shark or an elephant moving through the Rift Valley of Africa, and transform your size and peer into a cell nucleus, and so on. According to Bailenson (2018: 6), this is how you will feel and experience with VR:

> There is a major qualitative difference between VR and watching video. It feels real. Good VR does that. When done right, VR experiences—intense, beautiful, violent, touching, erotic, educational, or whatever else you choose them to be—will feel so realistic and immersive they will have the potential, similar to experiences in the real world, to enact profound and lasting changes in us.

The latter engendered condition raised by Bailenson (2018) of conducing a change in us having experienced the VR can be considered as either positive or negative to the individual person.

As an experiential tool, VR has the potential of educating the user with a "lived experience" of the costs of environmental damage that has been inflicted on the landscape such as logging the forest when we use soft toilet paper and tissue (Bailenson 2018). In a study done at Stanford University, a virtual chainsaw was incorporated in the experiment whereby once the subject entered the virtual forest, the subject would see his/her virtual arms holding a chain saw and when the subject approaches a virtual tree, the chain saw would roar to life and slice the tree trunk in a few minutes, and the noise of the tree trunk falling would be heard. Following the felling, the subject was free to look around the virtual forest now silent as the wild life had fled and the logged tree is on the ground. In this study, 50% of the subjects experienced the VR logging of the tree and the rest were just read an account of what it is like to be cutting down a tree. All the experimental subjects were given brochures with data about the consequences on the use of soft toilet paper and tissue, and the number of trees required to meet our consumption level. This is not the end of the experiment as the experimenter, without informing the subjects, in a setting 30 minutes later wanted to examine whether the experiment has impacted on the experimental subjects' conservation behavior. The experimenter in an experimental subterfuge would spill a glass of water as the experimental subjects were walking pass her. The experimental subjects would be asked to help in the cleanup and a stack of tissue paper was made available to them. All the experimental subjects helped in the cleanup. What is significant is that the experimental subjects who participated in the virtual tree cutting used 20% less napkins than those who were only read an account of what it would be like to chop down a tree. It seems that there is a noticeable increase in actual conservation behavior that VR can induce an experience that is more impactful on a person than other means of educating or communicating the issues facing us, such as via videos and brochures.

VR has also been utilized to increase awareness of the conditions at factory farms that produce meat such as pork and chicken for consumption sold over the counter in large-chains supermarket. Animal rights group such as the Last Chance for Animals have produced VR programs to meet their strategy of making consumers aware of the state and conditions the pigs and chickens are reared in. Viewers experienced the same experience as the animals. In contrast to these conditions, the viewer would also get to experience rescued pigs and chickens in an animal sanctuary. Hopefully, these experiences will force consumers to reconsider consuming meat or the least to demand that the animals on factory farms are reared in better conditions (Holslin 2015).

There have been other studies showing the impactful manner by which VR can provide us with "experiences" that can influence our perception of the actual reality, and thereby conducing us to make behavioral changes. For example, reducing the amount of energy that we use daily to meet our social needs, such as showering, and considering the energy needed to heat up the hot water for such an activity has been studied. Such studies with AI can then be linked to the impact of the provision of such energy and the effects it has on future global warming would be useful. Often, quantitative accounting measures of our social consumption can alert and make us more aware of the natural resources we consume, and hopefully conduce us to change our daily behavior.

Besides trying to reduce environmental damage, VR can be used to meet our needs; especially those living in advanced developed economies, to visit other parts of the world. In this way, it can reduce our footprint on the distressed landscape and oceans. There have been numerous attempts to protect the global landscape from the impact of tourism. Ecotourism being one of the concoctions developed. Despite the sensitivity to the landscape, ecotourism still utilizes energy and other natural resources needed for consumption by the ecotourist. Travel by airplanes and local transportation uses fuel, natural areas have to be developed for ecotourist accommodation, wildlife habitat and migrations are impacted besides the importation of goods for the ecotourist to consume that also requires energy. Tourism has increased tremendously over the years. For 2017, during the UN's International Year of Sustainable Tourism for Development, the number of international tourists' arrivals reach a total of 1.3 billion (Center for Responsible Travel 2018). About 17 years ago, the total of international tourist arrivals was only about 663 million (The International Ecotourism Society 2000). It has almost doubled over 17 years. These levels are not expected to drop. Like global population levels they always meet the projected levels and will always exceed the projections. Furthermore, apart from these international trips, there are another 5–6 billion domestic trips taken. Of the 1.3 billion of international tourist arrivals, it is estimated that between 50% of these arrivals are nature and wildlife related. This level of nature-related travel means the invasion of parks, natural habitats, coral reefs, and so on, continues. It has led to many calls for sustainable tourism.

With VR, the need to travel can be greatly reduced as VR tourism can also make it possible to appreciate the landscape and the wildlife. If it is watching of whales, elephants, rhinos, lions, zebras, hippopotamuses, giraffes, iguanas, sharks, and so on, a VR experience assures you of seeing what you wanted to experience because these beautiful animals will always be present in a VR. This cannot be guaranteed, for example, if you fly in on an ecotourist visit to Kenya or Botswana to see rhinos, lions, and so on. Of course, there will always be the purist who believes that a real experience cannot be substituted by a VR one.

Cutting down energy use in scarcity circumstances can be enhanced with the use of VR for meetings. In such a setting, social VR offers an opportunity, though not fully perfected yet, can be utilized to have the numerous meetings in the business world. In a global business environment, whereby management meetings require the participation of managers based on different regions to meet at the same locale for discussion and team building, social VR can provide the platform whereby these meetings can be held, and the quality of the interaction of the participants can be heightened. In the old format of video conferencing, all the participants are watching the speaker, but could not watch the reactions of all or some of the participants that an individual participant would like to if it were an actual meeting. In social VR, each participant with their avatars is able to experience that they are in the meeting room. Body language, facial expressions, eye gaze, and so on, if done correctly can be observed. The latter, as we know, is included in how we communicate and evaluate what the other person we are communicating with is telling us. This shift to VR meetings, if used extensively, is a great asset to reducing the amount of time and fuel utilized to travel to meetings, and the loss of worker productivity during business or duty travel.

In the previous chapter, we had discussed the need to increase the dematerialization process so that we will use fewer natural resources to meet our material needs. The advanced developed economies are based on a consumption pattern that encourages exponential increases. The amount of waste generated as an outcome of this consumption pattern is seen clearly in the waste dump sites. Can this endless consumption be reduced, especially those items that are really non-essential? Various websites on the Internet have been offering virtual items for consumers to purchase virtual goods, including virtual designer dresses. Consumers are buying virtual goods such as boats, planes, clothes for their avatars in cyberspaces, or for themselves so that their pictures can be taken in tjhese items and distributed via social media to their friends/followers. This cyber consumption has less impact on our natural environment and with the scarcity of resources that we are facing will mean less harm than conspicuous consumption that is occurring in the real material world. We should approach this as satisfying needs material-free. No doubt, there are some who will question the reality issue and the benefits that are associated with this.

The amount of human and financial resources and time required to produce the VR so that we can experience another setting is quite extensive. Teams of engineers and software programmers, storyboard specialists, actors, three-dimensional artists, and experts conversant with the setting are needed to reproduce the VR that an individual will experience. But once it is done, it can be uploaded to the clouds for access throughout the Internet for those with the equipment to experience it. The potential for it to be used as part of the educational materials for a Massive Open Online Course (MOOC) also exists. Carefully done, an environmental VR initiative to deal with an ecological issue can impact those who have experienced it and heighten environmental mobilization patterns.

AR, like VR, can as well provide functionalities that will aid in the varied environmental issues that we face, beyond what we have discussed in the previous pages. Increasingly, because of the costs to purchasing the AR glasses, the business sector has started to adopt the technology for its manufacturing and service operations. This adoption of AR in the business sphere can be seen in some identified areas whereby this digital advancement has been implemented. The Intel Corporation has utilized this technological capacity to guide and aid the recalibration of its chip manufacturing machines whereby its engineers do not have to be on site to undertake the modification. Using AR goggles, the engineer can guide his/her colleague at another Intel plant located in another continent to undertake such a task. Along a similar vein, GlobalFoundries Inc., another chipmaker has also used the AR technology to deal with repairs of the machines that it manufactures. In the past, it relied on traveling engineers to fulfill such tasks. The Volkswagen Corporation has used this technology to enable its engineers/technicians to guide the technician located in one of its Porsche dealerships to undertake repairs where direct human interactions are limited, especially when there is an ongoing disease pandemic prohibiting travel. In the area of health care, the use of AR to treat patients using Microsoft 3.06% HoloLens has also been done, especially during a time of disease pandemic whereby medical doctors can be shielded from possible infections. Environmentally, we can see that being able to complete work tasks over a wide space (between continents) without the need for travel can help to reduce the utilization of energy as well as the emission of greenhouse gases, notwithstanding the efficiency achieved.

The immense potential VR and AR can offer us in dealing with the environmental issues we are facing, and will continue to face, during this transition needs to be explored and discussed. The main issue will be whether there will be a net gain in terms of energy and natural resources savings will definitely be asked. Figuring this out will not be easy because of the interconnected effects of the planet's ecology. Using AI to assist us can be very effective. As an estimate of calculating what we will have in savings in energy utilization, if there is any, will be a start. As it stands right now, demand in global computing power from Internet connected devices, high resolution video streaming, emails, surveillance cameras, and smart TVs is said to be increasing 20% a year and consumes 3–5% of the world's electricity in 2015 (The Guardian, Environment Section, December 11, 2017). In 2014, the Internet had 75 million servers connected to it. It is expected that by 2025, the world's electricity consumption for the communications will reach 20%. Cisco Systems has projected that new Internet users will grow by 1 billion with 3 billion users in 2015 to 4.1 billion in 2020. From 2015 to 2020, global IP networks supported up to 10 billion new devices and connections, increasing from 16.3 billion in 2015 to 26 billion in 2020.

The level of users of the Internet from individuals, business corporations, NGOs, and political institutions clearly will continue to grow, even if what we have identified in terms of the potential of AI, VR, and AR for addressing our environmental concerns falls mostly on deaf ears. If this is the case, the infrastructure of connectivity and technological innovations are already in place whereby they can satisfy consumption needs so that planetary social, economic, and political order can be maintained during the global crisis scarcity conditions of this system transition.

Chapter 4

Cultural Reactions and Degrowth

Good Life for All?

INTRODUCTION

In our contemporary period, ecological crisis conditions inevitably generate social responses demanding the restoration and recuperation of the environment that has been degraded by centuries of global human exuberant consumption. By no means are these current social reactions new to the human enterprise. In the last Dark Ages that we all are more aware of, normally known as the European Dark Ages, centuries of capital and profit accumulation throughout the Iron Age also led to the rise of cultural and religious reactions to the ecological degradation. Such movements of social reaction during the last Dark Ages reached its heights during the Roman period because of the manner by which the Romans in their empire building and unceasing mission to bring Roman rule and lifestyle to the Eurasian world produced wide-scale environmental degradation, pollution, and landscape transformation in Europe and North Africa (Chew 2007, 2008). Within the Roman Empire, this Roman lifestyle of boundless excessive consumption stimulated the development of an environmental consciousness that propound socioeconomic directions and practices that are the very opposite of the then Roman excessiveness. As we have suggested in the earlier chapters of this book, this reoriented social consciousness (monasticism) by some Romans appeared first on the edges, and over time, as the crisis deepens, gained traction (Chew 2007, 2008).

In our current era of global crisis, are we also witnessing the emergence of such consciousness and socioeconomic practices that will realign our ecological orientation like during the Roman era? To repeat what we have stated in the previous pages, we do know that over world history, social ecological consciousness often emerges historically when ecological degradation and environmental changes reached extreme conditions (Chew 2001, 2007). The emergence of these challenges and recurrence of social ecological *mentalité* and awareness can also be considered as *reformist* moments (from an environmental point of view), for it can provide ideas and life practices that will veer us away from the directions that have generated the environmental crisis, and allow us to address the crisis with new frameworks for socioeconomic and political adjustments in order that the ongoing system transition we discussed in the previous chapters can continue without too much socioeconomic chaos.

SIMPLE ELEGANT LIFESTYLE LEADING TO AN ABUNDANT LIFE?

Economic social systems over the millennia have always exhibited unceasing growth and exuberant living lifestyles. In our modern era, whether the economic system is market-based (capitalist) or state-directed (socialist), the slogan has always been growth and progressive economic development for all, unequal distribution within and between nations notwithstanding. Modern economic social systems especially the developed exchange economies are characterized by the framework of an abundant life of objects for consumption regardless of whether these objects are of a material or non-material nature. In a similar fashion, those social systems that are state-directed such as the People's Republic of China have also attempted to follow the objective of commodities consumption as the overall orientation of organizing the socioeconomic system to meet social needs. Consumption is the order of the day!

Given such a principle of economic organization with unceasing growth and commodities consumption as the major objective tendency, it is no wonder that the stress on the natural resources of the planet will always be there regardless of the existing types of sociopolitical system to date. This means the call to end the "capitalist mode of socioeconomic organization" as some (Moore 2015, 2016; Robinson 2014) have demanded will not provide relief to the stress on Nature, other than perhaps a lower level of profits will be amassed by the owners of capital if the system stays the same. To progressive sociologists like Moore and Robinson, with their commitments to social justice, there is little possibility that the current mode of socioeconomic organization, especially with ongoing globalization, will satisfy their political objectives of the end of exploitation and class domination, and perhaps even for Moore (2015) environmental destruction.

Notwithstanding the above, we need to consider a different socioeconomic formulation. Numerous proposals to counter the existing growth model has been made during our current era, and lately under the umbrella of economic degrowth and the prospect for a sustainable society (Devall 1988, 1993, 2020; Chew 2020; Sessions 1995; D'Alisa et al. 2015; Kerschner et al. 2018). Do they offer prospects that will lead to a socioeconomic relationship with Nature that will be less exploitative than what exists during the current era, and offer models and patterns that perhaps will form the basis of reorganization of socioeconomic relations for the future?

AN ALTERNATE: BIOREGIONALISM

Starting with Christian monasticism and its emergence in the fourth century A.D. as a reaction to the turbulent conditions of the Dark Age of Antiquity reveals the response of certain segments of the Roman social community to the systemic crisis (Chew 2008). This response to the crisis that happened almost 1,700 years ago appears to recur throughout the history of Western civilizations as well whereby alternate worldviews and practices always re-emerge in response to the crisis conditions of the time. By no means is such a response to the systemic crisis restricted to just Western civilizations, the emergence of alternative views asserting a changing understanding of our relationship with Nature have also appeared in other civilizations throughout world history (Chew 2008; Whitaker 2009). Nonetheless in our current era, with the global crisis in diminishing natural resources, large-scale species extinction, incessant population growth and worldwide excessive urbanization, an alternate worldview and practice has reappeared in Western civilization. It had its start in the 1960s and 1970s of our current era and offers a counter to the dominant anthropocentric paradigm of unceasing growth and exuberant lifestyle (Devall 1988, 1993, 2020; Chew 2008, 2020; Sessions 1995). This *orientation* influenced by some Western philosophers such as John Locke (1891) and Spinoza (1941), suggests that harming other living things such as those belonging to the animal kingdom would be morally wrong, and that our natural system is composed of interrelationships with no hierarchical order, thus underscoring that everything is connected to everything else, and everything has the right to exist. Therefore, every living thing has its own worth. If this should be the case, such a philosophical understanding would be counter to the current basis of our socioeconomic and political understanding of our natural and social worlds based on the anthropocentric growth model whereby the human being stands supreme. There are numerous examples of environmental sociologists and other social scientists that adopt the above anthropocentric position from progressive scholars such as Moore (2015, 2016), Robinson (2014), Foster (1999a, 1999b, 2009), Gould et al. (2016), through to Dunlap (2002), and Mol and Sonnenfeld (2000) in their works.

In the Western world, in America, the Spinozan theme of interrelationships, for example, was taken up by Henry David Thoreau (1972, 1982). For Thoreau, there is an organic holism, whereby every living thing, for example, in the Maine woods, is a community of living things. Hierarchical order does not exist for Thoreau, and for him, every creature is better alive than dead (Thoreau 1972). Such a shift away from an anthropocentric bias to more of an ecocentric one was also shared by John Muir (1911, 1918, 1986). For Muir, everything in Nature has a value. The human does not hold court over Nature, but is just part of the living ecosystem. In this, Muir thus promoted the defense of the American wilderness. The continuation of this way of thinking and understanding can be found in the writings of Aldo Leopold, who insists that the human individual belongs to a biotic community of interdependent parts. From an initial utilitarian position toward Nature—the greatest good for the greatest number of people over the longest possible run—Leopold's belief transformed at the end toward viewing the human individual not as a conqueror of the land but as a plain member and citizen. Calling for a land ethic that treats the land and Nature with love and respect led Leopold (1949: 262) to suggest a land use decision is right "when it tends to preserve the integrity, stability, and beauty of the biotic community. It is wrong when it tends otherwise."

The above philosophical conception of Nature resonated with the various social movements that appeared on the socioeconomic and political scene in the early 1960s and 1970s in North America. The life practice proposed and attempted from this philosophical position would be counter to the dominant paradigm of growth, consumption, and industrialism. The 1960s and 1970s were turbulent times of protest over the excesses of federal-state control, the war in Vietnam, the corporate control of economic life, and the overall technologization of life in general. Such socioeconomic and political conditions mesh with the emerging counter alternative viewpoint. As a result, back-to-the-land initiatives to organize life that is in direct contrast to the orthodox practices then, and to return to Nature were the order calls of the day. A vision of life practices started to unfold and initiated the opportunity to further the development of the practice of what was termed then as the bioregional vision (Aberley 1998; Hansson 2003). Proponents of this worldview came from diverse disciplines and professions. Poet Gary Snyder (1974, 1980) wrote early treatises on the idea of bioregionalism. Peter Berg, actor and writer, of the Planet Drum Foundation was also another such proponent of the vision. Berg (1978, 2009) developed a manual on bioregional living that was widely followed. Others such as Jim Dodge (1981), Freeman House (1998), and Kirkpatrick Sale (1991) also added to this perspective. Jim Dodge retired in 2008 and is Professor Emeritus in the Department of English at Humboldt State University in Arcata, California. Freeman House was a former commercial fisherman, and a co-founder of the Mattole Watershed Salmon Support Group and the Mattole Restoration Council. The work of the Council was to restore the Mattole watershed, and to bring back the salmon runs.

What does this vision and life-practice cover? Bioregionalism's basic premise is that the human being is part of a living world, and as a result there should not be a separation between the human and other living things. This interrelationship between humans and their surroundings is considered harmonious, complex, and diverse—in short, a view of the world that is Gaean. It is in contrast to the twentieth century, and even today, whereby you find the following: a) excessive consumption of natural resources, b) the impending scarcity of these resources and the overdrawing of them by humans leading to species extinction, c) the uncontrollable urbanization, d) excessive deforestation, and e) climate change. What the bioregional vision calls for is a radical change in our *relationship* to our natural surroundings and our life practices.

Such a counter vision requires a different scale of life practice. Whereas the common or orthodox assumption of scale operates at the level of the state, and perhaps nation/global, the bioregional scale has region in place of state, and community replacing nation/world (Sale 1991). This shift in juridical-spatial dimension assumes that for bioregionalism, the emphasis is on localism instead of globalism that is commonly assumed then and continues to be encouraged even today. If the scale shifts to localism, then the focus is on the concept of *place* and how this informs our life practice (Aberley 1998; Carr 2004; Dodge 1981; House 1998; Sale 1991; Berg 1978; Berg and Dasmann 1978; Chew 1997, 2020; Devall 1998, 2020; Thayer 2003). Given this, the conduct of life practice means to bear responsibility to the locale where one inhabits. Such an obligation entails that essential life and natural processes of the place remain intact and evolving. In this sense, place is then considered to be the bioregion, with *bios* meaning life and *region* meaning life-territory (Dodge 1981). In such a context, life and place are intertwined (Thayer 2003).

By pursuing this sense of place, one also develops a direct relationship with one's immediate surroundings (Devall 2020). This relationship is then transformed into a set of life practices and awareness. It involves being aware of the rhythms and contours of the landscape and the ecological relations that underline its dynamics. It entails being aware of the diversity of life and the need either to restore what has been degraded or to ensure that ecological degradation does not take place. What it also means is the enriching of the sociocultural life and the natural aspects of the place.

To achieve this requires, the efforts of not only the development of an individual understanding and realization, but that of the whole community. The emphasis and responsibility then is on the community, and for the community to partake in such life practices over the duration of generations of community members. In this light, it is a call for a return to the community that has been forsaken lately for individual ego-centrality—the result of the manner our socioeconomic and political behaviors have evolved to date (Chew 2020).

This realignment of socioeconomic and political orientation to community redraws and realigns in spatial dimensions the parameters of a bioregion to be defined not by hard political boundaries that we have resided under such as state, country, or nation but to soft boundaries. These soft boundaries are to be determined by physical and biotic shift criteria such as changes in plant and animal life demarcating the boundaries of one bioregion with another (Dodge 1981). Coupled with landforms, such as watersheds and mountain ranges, would add to the demarcation boundary of a bioregion. Added to this, we can also consider cultural and phenomenological aspects of social life located within these soft boundaries.

If socioeconomic and political life are determined and defined by such boundary dimensions that are closely aligned to landscape and rhythms of natural life, the bioregion vision is shaped to *counter* what is viewed as sources and causes of the destruction of Nature, and the strain and tension of sociocultural relations. It is proposed as the counter alternative to our existing and practicing political economy and cultural activities. Sale (1991) and Carr (2004) offer an articulation of the bioregional vision in these spheres governing our social life. In the realm of economics, the bioregional view is that the economy is based on maintaining the natural instead of unsustainable excessive extraction of natural resources—in other words, an economy based on conservation. What this means is an economy dependent on "a *minimum* number of goods and the *minimum* amount of environmental disruption along with the *maximum* use of renewable resources and the *maximum* use of human labor and ingenuity" (Sale 1991: 69). In this view, maximal economic grow is not deemed to be the goal of such an economy, instead sustainability is preferred.

With conservation as an operating principle of a bioregional economy, self-sufficiency of the economy becomes the underlying principle. A self-sufficient bioregion would not only be economically stable, it would also mean that it will not have to be so dependent on other regions or other parts of the world to ensure its economic reproducibility. By avoiding this, a bioregional economy would not be destructive to natural areas and sociocultural lifestyles of other regions and parts of the world, as the self-sufficient bioregion would not be so dependent on these resources. With a self-sufficient bioregion based on the economic precepts listed above it would mean that all the issues and tragedies, and consequences of a growth-based economy ranging from pollution, diseases, toxicity, and crime could be avoided.

As the focus of this view is shifted to the level of community and place, in the area of governance, decentralization as opposed to centralization (a feature of our current global world) is the political norm. Where diversity of all life and interdependency relationships are celebrated, the emphasis is not on hierarchical efficiency. Rather, complementarity and reciprocity of relationships are fostered. What this entails is that there is no identification of hierarchical priority of functions or rank, for all parts are necessary to the reproduction of the whole. Shared responsibility becomes the norm of governance. With this form, governance, communication, and networking become necessary and important. Networks therefore ground the political governance to horizontal decision-making and autonomous units of political participation. Given the celebration of diversity of life, it would seem likely that different political relations would likely exist according to their own environmental settings and their own political needs. Instead of political homogenization that is the tendency of political globalization strategies today, the stress is the autonomous evolution of political practices that are compatible to life practices and the requirements of the cultural-economic landscape.

If this becomes the nature of political and economic governance, the sociocultural aspects of life would entail certain specific characteristics that embrace these thematics. The *principal of intrinsic right to evolve for all living things* suggests that the diversity of culture and other life forms is encouraged. With the linkage to place, the cultural understanding of historical relationship between the human community and the land is understood and remembered (Devall 2020). Cultural practices and transformation thus evolve and adapt to the biological integrities of the landscape.

Bioregionalism as the outcome to system crisis appears during periods when there are tremendous socioeconomic, political, and ecological upheavals. It provides an alternate worldview and life practices with an emphasis of returning to past interpretations of sociocultural life and Nature. These interpretations are not proponents of growth, consumption excesses, and centralization, for they are reactions to these tendencies that seemed to underline human societies over world history during certain crisis points.

Bioregionalism's historical development has yet to be completed. Offering a worldview and life practice that is different from what existed (and exists today), bioregionalism as a life practice has taken root only in isolated rural communities, and in certain pockets of urban settings. It has been adopted in a limited fashion in the public sphere of planning, land use, and renewable energy projects. For example, the California Resources Agency has divided the state of California into bioregions for planning purposes and in order to manage the natural resources of the state of California. Federal and state agencies responsible for natural resources and wildlife have also included the concept of bioregionalism and watershed distinction in their planning activities (Carr 2004). In urban centers such as San Francisco, Vancouver, and Toronto, bioregional groups have conducted civic projects with a bioregional vision in mind.

The bioregional vision of localism, self-sufficiency, place, and home resonates in an era of globalism that promotes exactly the opposite of what bioregionalism proposes. At this *conjoncture* of world history, the anticipation is that as the effects of globalism come to be felt further as history unfolds, the bioregional vision will perhaps offer a viable alternative to the globalization-growth model that is currently in play. Furthermore, with the current emphasis on consensus building as an approach toward governance, especially in the arena of natural resource use and planning, the possibility of cooptation of bioregional ideas and practices into mainstream environmental decision-making processes and procedures is very likely. What choices and practices we undertake will depend on the structural conditions of the planet as the millennium unfolds. From our past history, we realize that human choices are often decided due to crisis conditions, and that the selections we make, most often, are based on the contingencies we encounter.

ANOTHER RECENT PROPOSAL: DEGROWTH

More than five decades following the emergence of bioregionalism, the continuing global crisis conditions in the twenty-first century A.D. have once again precipitated a very similar style of theoretical reaction. This theoretical consideration with the signification of the word "Degrowth" shares the same argument that was first declared in the earlier Club of Rome reports (Meadows et al. 1972, 2004; Mesarovic and Pestel 1974) on the limits to growth. Degrowth is a theoretical viewpoint demanding the rejection of economic growth as a social objective (D'Alisa et al. 2015; Markantonatou 2013). If this is the case, there will be less utilization of natural resources leading to socioeconomic and political organizations that will be structured and shaped very differently. Hence, providing us with a different pattern of living.

For Degrowth proponents, the words such as simplicity, conviviality, care and commons, signify what the lifestyle will be like (Muraca 2013). Such a lifestyle, to borrow from sociologist Bill Devall's book (1993), is to live richly in an age of limits, though in a different manner. To do this, without sacrificing the quality of our life, we need to downscale production and consumption thereby reducing the throughput of natural resources and energy that underlines today's capitalist and socialist organized political and economic systems. It would, therefore, entail revisiting the formation of value in the reproduction of our socioeconomic institutions and agency. It means to analyze critically the circuit of value formation that is determined via the commodification process whereby ecological services, including land and social labor, are expended in the production of commodities finally arriving at a monetary value for exchange. To counter the contemporary circumstances under which these commodities are produced via capitalist and socialist production principles, the degrowth thematic also has proposed an array of alternative arrangements that a different route of commodification and value formation can occur that does not follow the principles of capitalist economic growth. Much similar to the bioregional perspective, they range from new forms of living and producing, such as eco communities and work cooperatives to basic and minimum income redistribution and sharing of work (D'Alisa et al. 2015). In a nutshell, a totally different pattern of socioeconomic and political arrangements that is green, caring, organized around a communal economy, much like what bioregionalism outlined almost five decades ago.

Rejection of economic growth as the basic organizing principle of human societies that has been the underlying basis of the social evolution of modern advanced industrial societies means that there has to be a revolutionary change in social cultural and knowledge systems that organizes these social systems. If growth and development are not the goals, then there has to be replacement ideas and values. Scarcity of natural resources, environmental deterioration, biodiversity crisis, and pollution has prompted a social and political questioning of the trajectory of the direction of social and natural evolutions that are taking place. Among the degrowth proponents, a panorama of ideas has been put forth that are the very opposite of economic growth theme (Kerschner et al. 2018; Buch-Hansen et al. 2019).

The degrowth initiative is to imagine and propose an alternative vision to the modern concept of development. This means it is not to be a model simply based on greener development like what the ecological modernization theories (Mol and Sonnenfeld 2000; Mol 2003) have proposed. What is this alternative to the concept of development? It means a transition to social formations that live simply with less but still flourishes. We can envisage the formation of eco-communities that utilize the principles of cooperatives, urban gardens, barter markets, and so on. Unlike contemporary capitalist production whereby value and profits are realized from market exchange, the rebalancing is toward use value. Thereby, there is a shift from exchange value to use value. Wage labor is deemphasized and transformed to become more voluntary activity. Social and market exchange are not geared toward maximization of profits but for the reciprocity of use needs circumscribed by the concept of the commons. The commons as a concept underline the principle of sharing and reciprocity in social relations.

Given the above socioeconomic circumstances, there will be unemployment that results. This by no means should be considered unusual in the transition process whereby the degrowth thematic is put in play. To alleviate such conditions, job guarantee schemes, and the provision of basic income to be financed from taxes can be the policy goals for implementation (Buch-Hansen et al. 2019). Work sharing can also be another viable alternative to pursue. The supplementation of state money and credit that is under the jurisdiction of private sector banks and federal reserve boards can be satisfied by community currencies—complements during the era of transition (Kallis et al. 2015). Such financial circumstances can provide the opportunities for debt review and forgiveness since growth no longer forms the basis for societal reproduction.

The above broad contours of the Degrowth scheme, and what has been produced and debated so far included the various alternative movements and ideas from the 1960s counter culture, back to the land, alternative living, movements, and so on (see, for example, D'Alisa et al. 2015). From community currencies to urban gardening that articulates a different perspective to social living and lifestyles, there is the veering away from growth and profit maximization, besides a strong support for decentralized systems. Will these recent Degrowth ideas and proposals take hold and gain momentum? That depends on the conditions of the time. Like Bioregionalism, it has emerged during a difficult period of ecological, socioeconomic, and political stresses, and has offered a different alternative to current existing capitalist and socialist practices.

FOOD AS LIFE: COMMUNITY-SUPPORTED AGRICULTURE

Bioregionalism and Degrowth have provided a panorama of models and social practices that can provide possible solutions to meet the system crisis limits that we are experiencing globally. Food production that forms the basis of human life perhaps is the most promising avenue for us to explore the bioregional and degrowth ideas for suggested agricultural practices. In contrast to industrial agriculture that predominates our food-producing practices that have led to soil infertility and erosion, water contamination through pesticide use, biodiversity and water table losses; the practice of community-supported agriculture (CSA) can be an answer during an era of severe system limits as a consequence of the effects of global warming (Kimbrell 2002). Given that more sustainable agricultural practices have to be undertaken in view of the growing demand for food with world population expected to grow from the present 7.5 billion to 8.5–9 billion by 2030, innovative practices that do not repeat the problems industrial agriculture have generated need to be considered (Alexandratos and Bruinsma 2012).

Besides developing a different model for growing food such as CSA, consideration for changes to be made should be directed beyond the farm to the food delivery and exchange structures that are now in place in advanced industrial societies where consumers accessed the food distribution system via supermarkets that have food delivered across vast distances. The transportation distances could cover between regions of the world during seasonal periods whereby a particular food item due to climatic conditions cannot be grown. For example, the sale and consumption of strawberries during winter in the northeastern part of the United States during a period when the fruit cannot be grown. Most often, the fruit will be shipped for sale from California, Central America, parts of Latin America, and North Africa. The other issue to be discussed is whether the consumer should eat a menu that is based on the growing season of where the individual resides, instead of choosing a menu that is based on any season regardless of climatic growing conditions. Choices made will impact on energy utilization in the area of food consumption.

Food and its consumption are the most important aspect of our social formation for we are materialistic beings and we need it to reproduce our lives. The impact of this daily activity flows through from the micro to the macro levels of our political and economic social structures. At the macro level, it determines the structure and processes of our economy, policies guiding our land use, and the distribution of profits and income. For the latter, value realized is not only in terms of monetary terms, but also the quality of lifestyle and health for the entire population. At the same time, this translates to the micro level of the individual and the family. In light of all these translatable impacts any structural change will have tremendous impact on the dynamics and operations of the social formation. In this regard, should there be a change in the worldview of food production, delivery and consumption practices in reaction to the increasing awareness of the various negative and degradative facets of industrial food production, we can witness a socioeconomic transformation and challenge to existing exchange relationships, and the individual choices that are made. The current overwhelming interest by individual consumers and their families in purchasing food that are organically grown is a case in point.

The increasing awareness of the quality of food produced by agribusiness has, in certain cases, prompted at the level of the family household to seek alternative sources of production for the food that it consumed. One such socioeconomic arrangement is the CSA. Simply put, this socioeconomic food production and relationship structure is defined as the growing of local food for local people at a fair price along with a fair wage for the grower of the food, and having the activity carried out in a manner that is sustainable to the land. So far, CSAs exist mostly in rural communities.

By the late 1990s in the United States and Canada, there were approximately over 1000 CSA farms (Bramble 2009). This farming arrangement is organized in four forms (Henderson and Van En 1999). The first type is a subscription or farmer-driven model, whereby the farming arrangement is organized by the farmer with some members (shareholder) of the community belonging to the farm via a subscription payment (farm share) for each growing season. In this arrangement, the contribution of the shareholder is mainly economic with some opportunity for the shareholder to work on the farm if the shareholder wishes to do so. The second type of CSA is one whereby interested community members band together and employ a specific farmer to grow the crops for them. In this case, the CSA is shareholder driven. The third type is the farmer cooperative. In this arrangement, two or more farmers pool their resources to grow a wider variety of food products. The fourth type of arrangement is the farmer consumer cooperative with the farmer and the members of the community co-owning the land and working together to grow the crops.

CSA farm sizes are approximately 35 acres, producing about 5 to 10 pounds of food per week enough for two to three adults with the number of shareholders varying depending on the acreage of the farm. Cost-wise, a share in one of the CSA farming arrangements is about \$10–\$35 per week for an average growing season of twenty-two weeks. The cost per share rises if the food produced is meat products. An average share for a member would average about \$500 per growing season. Selected studies have suggested some cost savings for CSA shareholders in comparison to the equivalent amount of food items purchased at local supermarkets (Dyck 1992).

What is significant about the CSA arrangement is the terms of the socio-economic relation. The consumer signs an agreement with a local farmer and enters into a contract by agreeing to pay a monthly or seasonal charge for a weekly basket of fruits, and vegetables grown by the farmer. The food products could also be meat and meat products. By such an arrangement, the farmer is provided a level of financial security that his/her produce will be exchanged thus reducing the farmer's risks, unlike traditional farming whereby the farmer assumes all the financial risks. Such a transaction means that the farmer's crops are sold before they are even grown.

This CSA's model also impacts on the exchange value of the farm products as there is no longer any distributing and middle-person costs. Energy consumption is reduced as the produce are distributed and eaten locally, unlike industrial and agribusiness produced food that travel usually long distances to the market. Studies have shown the distance traveled. For example, an average carrot is transported over 1,838 miles from farm to market (Bramble 2009). In a globalized world of today, the distance traveled for certain food items is even further. With interconnected world of global air and sea transportation, food and meat produce are transported between continents. To be able to offer seasonal food items all year round to meet exuberant consumption, food items are shipped by air if they are perishable, thus increasing energy consumption and global warming.

Consumption of local food that are grown organically using sustainable farming methods has increasingly being sought after these days because of its nutritional value for our health, notwithstanding the taste. Farm to fork is increasingly the order of the day, especially for the upper elites of advanced industrial societies. Notwithstanding this, the norm of the present time is the consumption of organically produced pesticide-free food, and this "mantra" has settled into the minds of the upper strata population of advanced industrial economies. In view of this, the large-scale private business enterprises have jumped into this niche, used to be occupied by food store cooperatives, where we now see the exponential growth of organic style food outlets such as Whole Foods stores in urban areas of the United States. There is no doubt that this move will continue to intensify and expand as long as profits can be made in tandem with changing consumer purchases and tastes.

It is clear that the CSA model does reduce energy consumption and does not impact negatively on the environment like agribusiness does, since the food produced and exchanged do not travel far—the 50/500-mile menu. Large-scale organic food outlets, such as Whole Foods—now owned by diversified conglomerate Amazon—though might have less impact on the environment, continue to consume energy and other natural resources as they continue to repeat the distribution and exchange processes similar to the standard supermarket model that we are fully conversant with.

Along with the CSA model, the distribution and exchange places for farm produce are further enhanced by the formation of a structure called the Farmer's Market. Farmers' Markets provide the opportunities for CSAs and small farmers in mainly rural areas that are close to larger urban environments to sell their farm produce to the urban dwellers. The number of such markets in the United States was 7,175 in 2011, and their growth has been phenomenal when in 2005 there were only 4,093 in number (Zezima 2011). With the increasing awareness of healthy living and the health dangers with the consumption of industrially grown food, sales at these Farmers' Markets have been booming initially. Such levels of sales, and with the increasing number of Farmers' Markets have led to the closure of some (Zezima 2011).

Besides offering a locale for locally produced food to be available for sale, the Farmers' Markets also provide an opportunity for the local population to have an opportunity to meet their neighbors, and a place for local artisans to sell their locally produced goods that are not specifically food items. In some ways, it is similar to attending Sunday church services for rural communities at the turn of the twentieth century whereby following the religious service a shared meal is enjoyed and communal exchange of social news and happenings.

How then can we use the CSA model to scale up to meet the food needs of a growing world's population, so that more can enjoy the bounties of such a model? The CSA's farming practices and structures can be enhanced by the AI of learning machines that have the capabilities and capacities to generate learning algorithms that will streamline, enhance, and modulate the efficient flow of energy and other inputs, thus providing strategies and directions of the efficient and effective use of geographic space and time. The inclusion of small robots that can monitor the size and health of the plants that measures the stem diameter, leaf-area index, and stand count can assist the farmer to devote his resources to increase yield per acre. All these initiatives are to be guided by the changed socioeconomic relation whereby the consumer with a farm share in the production process ensures a reduction of risk to the farmer. Thus, the exchange value for the food at the market will be reduced. Membership in this type of socioeconomic relation such as the CSA underlines the potential for the resurgence of the concept of community that has so far been fragmented by market relations in a post-industrial era.

Participation in a CSA arrangement will also lead to the member (if s/he chooses to) being an active player in the production of the farm produce that is part of the responsibility of the socioeconomic exchange relationship. In this way, the members and their families are exposed to the many facets and knowledge related to growing of food, a sphere of life that so often is foreign to urban dwellers. In this regard, the distantiation from Nature that distinguishes urban living thus can be reduced should the model of the CSA be increasingly adopted as part and parcel of daily life.

System transition, as we have mentioned in various parts of this book, can force us to re-examine the structures and processes that have delineated the manner we have reproduced our lives especially when we are facing various crisis conditions. Ideas and social practices often emerge to deal with the systemic crises. The adoption and experimentation that ensue will take time and also transcend geographic space especially in a globalized interconnected world. Information and knowledge on how to grow food in different socioeconomic relations can no longer be localized with the existence of the Internet and other electronic media.

Chapter 5

Living in an Era of Scarcity

INTRODUCTION

Bill McKibben in his book *Falter* warned us about the threat that global warming poses to the survival of living things including humans on planet Earth. Qualitatively, in a nutshell, he identifies and surveys globally the impending death of our natural environment. The ringing alarm he has sounded by no means is new, except this time the source for the alarm comes from global warming gases such as carbon dioxide and methane, whereas in the early 1990s, global deforestation and the erosion of the ozone layer were the dangers. The clarion call for all hands-on deck to face this threat by environmental activists such as McKibben, needs to be complimented, especially for their energies devoted to showing what global warming can inflict on all our lives, and the social chaos that follows. From this angle, environmental social movements can act as agents of change to transform the socioeconomic and political structures in place that have created the present conditions. Whether such activist projects can lead to successful transformations in socioeconomic conduct and policies remain to be seen and realized. In the meantime, however, should we rely only on the "agency" of the environmental movements as the only avenue to pursue socioeconomic structural change to deal with the threat, or should we also consider other structural analyses that might prevent the chaos, and provide pragmatic solutions that can meet the reproductive needs of the social system in crisis?

CYBERSPACE AND DIGITALIZATION

It is very clear that dense human occupation has taken place in most of the geographic spaces on this planet apart from the Arctic and Antarctic regions. In other words, the world is getting very crowded. More people now live in cities than in rural areas—a first in our human history (WBGU 2011). Prior to this, at the start of the last century, only 10%–15% lived in the urban areas. What this means is that three and half billion people now are in urban areas with the global population projected to rise from 7.5 billion to 9 billion by 2050 (WBGU 2011; Markantonatou 2013). Such a trend suggests that by 2050, at least 2 more billion persons will be found in urban areas of the world. As urban sociologists and planners have stated, urban dwellers, except the homeless, tend to consume more energy and resources than those in their surrounding rural areas. We can also expect that with increasing affluence, the global consumption levels across all regions of the world will continue to trend even higher. To this extent, the urban areas of the world now account for 75% of the world's energy consumption. Even though the expected increase in urban population will be in the less industrialized countries, this does not mean that the consumption level will not continue to move upwards, albeit not as drastically. But if we add to this the consumption level of advanced industrial countries, it means that the trend cannot be changed much as the latter continues to consume about 50% of the global energy but having only 20% of the world's population (WGBU 2011).

Given the above statistical trends, we must realize that the global economy cannot afford to stagnate. What this means is that the dominant economic relationship on this planet depicted by *commodity exchange* between individuals, groups, nations, and regions, *whether determined by the market or State*, will require continued expansion, recalibration, reconceptualization, and perhaps, redistribution so that social order can still be maintained or else global crisis will likely follow.

To repeat, even with the distribution of the global surplus being uneven and unequal, economic growth and expansion will continue in the short and medium terms. Specifically, if the effects of global warming occur on schedule, *expansion and growth* of economic market exchanges will need to continue unabated in light of the exponential growth of the world's population if not global social order will become chaotic, and maintaining it will prove to be difficult. Global disruption will be the order of the day. Putting aside for the time being the issue of striving for the good and the just society proclaimed by modernity and the progressive Left, there is a need to consider wisely the decisions that we have to make and foster in light of these circumstances so that natural and social evolution can persist.

I am sure we all hope that social and natural evolution will continue, and if this is the case, social adaptation to the changing environment will need to be pursued and fostered. In the previous chapters, I have identified some structural tendencies that have emerged in the arena of technological innovations and development in cyberspace giving rise to a digitalized world that offers significant potential to address issues that we face in natural resource scarcity, deforestation, biodiversity losses, and global warming. Furthermore, emerging ideologies and changing social relations derived from the net economy through digitalized products can provide different structural pathways to overcome the environmental issues and concerns. Digitalization has offered us numerous potentials in using Big Data, deep learning machines, virtual/augmented realities, and artificial intelligence in this new millennium to respond to the various global crises and the transition we now face.

Global transitions are rare. They occur when a social system is beset with crises that undermine the continued functioning of the system that some have stated that we are now facing. Transitions, therefore, can be viewed as phases depicted by the process of de-structuration and restructuration of the overall structural framework of the organizations of the social system. Transitions, or what I have categorized as Dark Ages, by no means are smooth and chaos free (Chew 2007, 2008). Lengthy periods of twists and turns will occur. Recalibration and reconceptualization will need to take place for ideas, values and strategies to mesh with the new technological base such as what digitalization can fundamentally provide to our dated structures and socioeconomic and political processes that no longer cohere. With the evolving socioeconomic and political structures and processes, there might be the opportunity for what digitalization has provided initially in terms of speed and efficiency leading to a changed marketplace fundamentally in terms of competition and incentives. Addressing issues of scarcity in material resources by developing processes and technologies that can handle these contingencies will perhaps deal with the environmental pressures we all face. Different players can enter the marketplace. Such developments can also mean that potentially the distribution of value creation can be changed from what it is now. The potential redistribution will mean that different segments of our society that participate in the production, distribution, and exchange of goods and services will have the opportunities to have either a larger or smaller share of the value (e.g., profits) of the products exchanged in the marketplace. The success of this redistribution will depend on the key players participating. This all will also depend on how successful the environmental social movements, the socioeconomic and political elites have been to change the hierarchical distribution of power, the level of scarcity in terms of available natural resources, and the pricing that results affecting the terms of valuation. There is no guarantee, as world history has shown, that those at the bottom of the hierarchical ladder will move up. In this respect, with regards to natural resource inputs in the production cycle, Moore's (2015) world ecology thesis of "cheap" Nature that enhances capital accumulation no longer holds at this current stage of world history.

Instead of focusing on a singularity issue such as global warming that is quite popular and in vogue these days, we need to re-conceptualize and recalibrate the structural processes that have been in place in our work and social routines. Since the start of 1970 when the microprocessor was introduced, we find the emergence of digital technologies with its digital infrastructures, products and services that enable different ways to meet work routines, personal living styles, wants, and needs, and new and different ways of conducting our work and personal lives (Chew 2008; Castells 1996, 1997, 1998). From such beginnings, a cyberspace distinguished by its virtuality has provided us with a social, economic and political reality that we have not envisaged earlier on that we now take mostly for granted. Increasingly we are comfortable with a non-materialistic world of commodities. As we know, commodities and their consumption (in a broader context) have underlined market exchanges for the last five thousand years at least. Such a marketplace or *agora* has existed in relationship with Nature, for it relies on the provision of natural resources of Nature for the market place to continue, and for socioeconomic and political life to be reproduced. This reliance on material resources for commodities production is increasingly being reduced by digitalization when manufacturing processes are based more on dematerialization and miniaturization or that commodities produced are increasingly material-free as we have highlighted in the previous chapters of this book. This trend will continue with increasing computing power. According to scientist, Ray Kurzweil of Google, by 2020, a thousand-dollar PC will have the computing capacity of a human brain that can make twenty million billion calculations a second (McKibben 2019: 136). By 2055, according to him, $1,000 worth of computing power will equal the processing power of all the humans on the planet, and by 2099, a penny's worth of computing power will be a billion times as powerful as all the human brains now on the planet (McKibben 2019: 136). Such computation power also raises future dangers that we cannot foresee.

Faced with natural resource scarcity and exponential population increases, and the potential of what digitalization can provide, we need to shift to new models of conducting and managing businesses, producing our food, organizing the market place where we conduct our exchanges, and designing the conduct of our leisure and sporting activities. The advances in digitalization can help by providing us with tools that will enable us to come up with a "new" model of thinking so that we can survive the transition and the anticipated chaos that so many experts have warned us. The potential of deep learning machines, artificial intelligence, virtual and augmented realities that we have discussed in the previous chapters can open new vistas to reproduce our lives in an increasingly scarce and crowded planet. We can consider these recent developments as giving us the opportunity to think "wisely" that in ancient cultures only occur when a person has had a lifetime of worldly experiences.

MANAGING BUSINESS AND SOCIAL REPRODUCTION

With increasing global population growth and the scarcity of natural resources, management activities reproducing our social and economic life needs to develop new structures, systems and processes that are more sustainable in the long run. Undertaking the digitalization of all records, forms, information, plans, and so on will create a more sustainable environment as little or lesser material resources are then needed to hold or store these records. New digitalized infrastructures will come into existence that will use sustainable resources and energy to replace the dated way we have been conducting these business and social reproduction of our lives. For example, the electronic transfers of information in these arenas will mean reduced use of any materialistic commodity such as paper, thereby reducing the need to rely on the trees or energy recycling process to manufacture paper. In a connected ecosystem, this shift will mean less heat trapping gases being emitted with fewer natural resources required to undertake these activities.

Other administrative activities in the business sector have benefited through the use of AI apps that with deep learning machines can perform large scale analysis of business transactions such as in the area of auditing that normally will mean laborious and lengthy time periods to meet administrative auditing requirements for publicly listed companies. Artificial Intelligence can perform the tasks of cross correlations, analysis of unusual patterns of transactional behaviors and provide choices and strategies to meet corporate goals. Its use, according to a recent study prepared by Pricewater Coopers LLP, and reported in *Wall Street Journal* article, *How to Support Widespread Adoption of AI*, artificial intelligence has the potential of boosting global economic output by between $13 and $15 trillion between now and 2030 (Wladawsky-Berger 2019). What this means in terms of savings *in material and energy resources is yet to be known and calculated accurately nor the problems AI can generate through its use needs to be studied* (Smil 2014).

Furthermore, with global business increasingly being the norm of business operations, the conference meetings that are required to coordinate business operations and collective decision making will benefit from meetings that utilize augmented reality teleconferences whereby the meeting participants do not physically travel globally for business meetings and conferences. They can send their avatars to such AR designed situations, and because of the increasing development in AR, the quality of the participant interactions is much more positive than current existing technologies such as using Skype or Zoom. With the increasing research foray in *affective* AI, such development can be harnessed into the meeting interaction that will further enhance and enrich the participants meeting interactions.

Having AI that has the capacity to recognize and respond to emotions, moods, and facial expressions and undertones will increasingly expand the capacity of the use of AI. As it stands now, these skills though present in all humans are absent in chatbots, digital assistants, other machine systems, and robots. The development of such software is being undertaken by technological corporations such as Google, Microsoft, and Affectiva. The aim in the long run is to have a seamless interaction between humans and machines. Beyond the incorporation of such systems in meeting environments, the emotional awareness developed in AI can generate many possibilities in helping us in our lives. For example, if your digital assistant recognizing your mood when you speak into your cell phone can say things to you to calm you down if the health monitoring feature in your cell phone are indicating to it of health dangers, or your GPS in your car senses that you are tired through facial recognition, you are alerted to stop the car and pull over to rest. Affectiva has over 1400 clients using its affective AI to gauge consumer reactions to advertisements through webcams. Affectiva's emotion recognition software through machine learning can track expressions across gender, race, and other demographics that appear globally.

Meeting preparations can also be facilitated, and perhaps, made more efficient and less stressful with the use of AI. These days with the norm in the business environment of emphasizing team coordination and networking; the predominance of presentations in meetings using power-points has been the *de rigueur* of meeting protocol and modality. The availability of an AI powered presentation coach launched recently by Microsoft Corporation is to ease meeting presentation preparation. Microsoft's meeting presentation coach was developed to help presenters who have the tendency to speak too fast or in monotones to enhance their speaking delivery. The user of the AI software can prepare for the user's presentation by speaking into the user's laptop, desktop, or smartphone, with the AI software providing real-time feedback on pacing, word choice, and other speaker's tics such as "anyway," "you know," and so on. This coaching is carried out with the software producing an instant report after the user has finished practice speaking with suggestions on areas for improvement. According to Microsoft, the AI software is listening to the user's speech and analyzing the words powered by the AI's voice recognition capability to convert natural speech to text and searching a database of flagged terms, and what a human public speaking coach would normally suggest. The software-database will be further enhanced when the user revises or ignores the software's suggestions through multiple practice sessions. In other words, through machine learning the coaching provided by the AI software becomes even more applicative for the user.

Other IT companies, such as Cisco Systems, have also introduced smart technology with tools that can identify participants' background attending a meeting using Cisco's AI software. This AI software has facial recognition capability that will automatically scan the Internet for biographical information of meeting attendees, thereby, instantly, informing the chair or the speaker of the meeting the background information of attendees so that the meeting can be more productive and engaging.

The increasing advances in AI have led to the development of AI systems that can complete investment memos, produce stories, and even write letters following the style of famous persons. This type of AI driven software is at the beta stage of testing and is named Generative Pre-Trained Transformer (GPT-3). When available, it will generate major changes in the work of news reporters, lawyers, and those involved in work requiring cognitive skills involving plain English. The end result, for example, in a lawyer's work, it will mean that instead of drafting the written text, such as legal documents, the lawyer will spend his/her time editing the documentation instead (Price, *Wall Street Journal*, August 22, 2020).

Such innovation indicated comes under what is termed as smart tools and robotic process automation technology. This area has been adopted by numerous companies and has reached a global spending of $680 million in 2018, and will soon reach $2.4 billion by 2022. It is expected that, in the future, some form of this technology will be adopted by more than 80% of the companies (that gross $1 billion in revenue) globally.

The possibilities are endless with the use of AI. The above are just examples of what enhanced digitalization can provide in the business environment and in our daily living. What we need to realize is the amount of materials and energy that we can save and realize, especially in this context; carbon credits do not have to be purchased for air and land travel to these meetings like what Al Gore has recommended for meeting environmental correctness!

With digital technologies we can explore new and alternative business models such as those fostering green digital start-ups and cooperatives. Geographic spaces can be transcended, that, in the past, have prevented collaboration that can lead to collectively organized economic systems oriented to the common good. Digitalization can allow for participatory value creation via the concept of a sharing economy that is also being promoted by the Degrowth movement that we have discussed in the previous chapter.

The vast expansion of online commerce has generated both positive and negative effects. The business operations of the global titan of Internet commerce, Amazon, is a case in point. It has been reported in 2019 that Amazon ships about 5 million packages a day, and its shipping costs have grown from $5.5 billion to $61.7 billion from 2010 to 2018. Notwithstanding the scale of business operations, this type of business activity produces positive outcomes in terms of vast savings in energy costs if we were to compare individual purchasing trips to reproduce the lifestyles that we foster and are accustomed to. Logistics that are involved in undertaking the reproduction of all our needs are streamlined and made more efficiently sustainable. Furthermore, the lesser need for warehouses and storage spaces for each individual businesses will free up existing defunct stores and mall spaces (brick and mortar retailing), and provide the possibility of turning these urban distressed mall areas back into green spaces instead of transforming them now into health clubs, and so on. The functionality of Amazon's services is realized further with the lockdowns of businesses during a disease pandemic whereby consumers can access their socioeconomic reproductive needs via home delivery.

Negative outcomes of the rise of online commerce—besides the issues of labor rights—cover as well unnecessary packaging waste and delivery services if there is little coordination in the latter activity. Packaging wastes have been reduced by innovations in packing and shipping. The volume of U.S. cardboard used was lower in 2015 than they were in 1995 (McAfee 2018). These issues raised can be either solved or reduced through management logistical studies that utilized artificial intelligence to correlate Big Data and prescribe solutions that are not only efficient but also sustainable. Algorithms and digitalized instruments can be developed using machine learning to direct and fine-tune the structures and processes.

AGRICULTURE AND FOOD PRODUCTION

With increases in the world's population and climate warming leading to threats to our food production system, there is a major need to ensure land use be planned and acted upon accordingly for the survival of the Planet. Even with the prevalence of industrial agriculture, there needs to be consideration that that food production be aimed for sustainability. In other words, from farm to fork regardless of whether it is industrial agriculture undertaken by corporations or small-scale agriculture, the key is to address the sustainability of the farming process.

Reduction of the use of pesticide and fertilizers to lower the environmental damage needs to be conducted across the spectrum of farms producing food, especially the corporate-sized farming operations. To do this, digitalization along such lines as the utilization of satellite imaging, drone surveys of the conditions of the farming land, irrigation areas and frequency are available and utilized in farming. This ensures the precision of the farming inputs used to maximize the necessary outputs for the marketplace. Precision application of fertilizers and pesticides to specific plants or trees could potentially reduce the use of these products by up to 80% (Ford 2015: 25). It can also reduce the run-offs into streams and rivers, thus ending not adding to the pollution of the water drainage systems, that in turn could affect the population of the fisheries; and thereby reducing the threat to the fishing industry.

Utilization of drone surveillance to improve efficiency and enhance agricultural production outcomes, as expected, has been employed quite extensively in agricultural activities involving agribusiness corporations. Drone development and utilization have been geographically extensive globally. The use of drones in palm oil production in Malaysia to deal with bagworms infestation of palm oil trees is a case in point. With the utilization of drones that enable the opportunity to target specific areas of such infestation has not only reduced the amount of pesticide use but has also reduced the costs of the palm oil production. Usual practice to deal with bagworms infestation is to employ piloted aircraft that not only costs more but also the application process is untargeted, hence a higher volume of pesticide use. According to Malaysian-based Braintree Technologies SDN, that has developed these drones, the use of such technology has increased yields by 20% and savings of as much as 50% of the costs of the oil palm production process. The adoption of this drone technology does not excuse the practices of deforesting virgin tropical rainforests by such countries as Malaysia and Indonesia to have a palm oil producing industry whereby the virgin forests are sacrificed to produce palm oil as an input in the manufacture of food commodities that are mostly consumed in the developed countries.

Another aspect of the digitalization process that is steadily gaining momentum is the collection of vast amounts of data (Big Data) that are sponsored by agribusiness corporations such as Mosanto and John Deere (Kelly and Knezevic 2016). Farmers in North America are encouraged by agribusiness to collect farm data such as soil conditions, weed varieties, and weather housed in a data platform such as Integrated Field Systems, Farm Forward and Weed I.D. so that their farming efforts can be maximized and risks reduced. Of course, the Big Data collected can also be used by the agribusiness corporations to pinpoint and target which type of pesticide to manufacture and where to market their weed killers in light of data on weather conditions. *Whether the profits gain from such technological innovation will be distributed is yet to be confirmed. Notwithstanding this, there is a global need to increase food production in light of the impending threat to food production globally from expected climate changes*. Areas of the world such as in the northern hemisphere that in the past would not be conducive to food production will be available if the temperature increases occur as predicted.

Besides large farming operations, digitalization can also improve the efficiency and productivity of small farms as well, especially those in the less advanced economies, by providing access to information and advice from extension workers. This latter case is important as agriculture in the less developed economies is often farmed on small plots of land, and with very little capital investments to increase the harvest. It is important to realize that the fastest rate of population growth is occurring in what is now defined as the South. Where the water is scarce and ecosystems are stretched, the challenge to find cheaper modalities of increasing the harvest is not only necessary but is also critical. Notwithstanding the harvest losses that occur in the postproduction systems (PPS) in the South, any help to increase harvest yield would be welcomed.

The emerging interest (such as digital food activism) demanding to know what we eat in terms of how the food is grown through to its supply chain cycle to reach the market place requires a level of transparency from farm to fork that will impact on its final sale and consumption (see, for example, Schneider et al. 2018). This is especially demanded by consumers in advanced market economies with the wide-scale dissemination of food consumption information and related health dangers and issues. Food Justice demands concerning access, equity, safety, and transparency will increasingly hold sway in view of the level of digitalization and social media available today (Caldwell 2018; Rousseau 2012). In this context, Food Justice also addresses accessibility to food information that is not always available to everyone, as there are those in the less privileged groups who do not have easy access to the Internet.

Digitalization can now provide the capacity to track this for the consumer's information. It can also be utilized to target blame by the health inspectors should a health incident occur that is caused by the grower, processor, packager, or truck driver. Tracking systems have been introduced to monitor and regulate food at different moments as it goes through the food production system. Crops and animals are tagged and entered into databases for tracking as they moved through the food system. For example, in more detail, a developed digital app that will be in use by 2020 can track the lifecycle of the specific chicken from the farm through to the market. Whether the specific chicken lived a free-range life on the farm can be tracked as each chicken is attached a tracking device on its foot that automatically uploads in real-time to a digital ledger or block chain. Sensors monitor the environment in which the chicken is raised such as temperature and humidity, and AI evaluates the chicken's health through the monitor sensors.

Such level of monitoring increases transparency and can also provide information concerning food safety issues should there be a need. The eco-conscious consumer can now be more assured of the quality of food s/he consumes, especially with disease outbreaks, food safety inspection lapses, concerns over contamination, and animal welfare. For example, for the latter concern, in terms of enhancing animal welfare, the dairy industry has a rather sophisticated system in monitoring the dairy cattle (Caldwell 2018). The dairy cows are tagged specifically with electronic chips and scanned by sensors that restrict them to go into each individualized assigned milking stalls. Once inside these individualized stalls, the sensors can then measure their milk output and the amount of grain that each of the milk cows has eaten. It, therefore, enables the farmer to have better oversight of the cows' health as the farmer can increase the cows' intake if they are slim in weight. The cows also have the ability to approach the individualized stalls if they are hungry, and the cows have the freedom and agency to enter the individualized stalls for extra feeding.

Such level of monitoring and availability of information can add to the valuation of the food product consumed and as well, increase the profits of some of the participants in the commodity value cycle. The distribution of value creation in view of such practices can also change. In this case, it might be to parties (such as corporate agriculture) that already have a larger share in the value distribution. But in some farming models that I have elaborated on in the previous chapter, the distribution of value creation might not necessarily be in this direction. Community Based Agriculture is one such model that we have identified whereby the distribution of value return has the tendency to be more egalitarian. The consumer, in this case, has a farm share in which the farmer after calculating the food growing costs determines the value of each farm share to be paid. In this manner, the exchange value of the farm share does not have to include transportation costs (the farm shareholder picks up the weekly share), storage and market space (supermarket) costs. In this context, value creation and its distribution are somewhat reallocated than the standard orthodox model of food production, distribution, and consumption. Furthermore, it will not need a social protest movement to achieve value redistribution that the radical left has clamored for a return of "class struggle" back into the food production (Nature) (see, for example, Leonardi 2019).

DIGITALIZATION AND ITS POTENTIAL CONSEQUENCES

Few can deny that the digital revolution has impacted immensely on our lifestyles and work. The ubiquity of the things produced from this revolution and utilized by everyone in their lives would be endless to list (Rubin 2018; Scoble and Israel 2017). There are carpers as well. With the capacity of artificial intelligence that can collect, pool, and analyze large amount of data in both the public and private sectors, the issue of intrusion of our individual privacy and rights have been raised (Davenport 2018). Notwithstanding this, the digital storage and pooling of (big) data in server farms, and in the cloud means that illegal and legal (for profit) intrusion in these data collection centers have also been raised. This then extends to the issue as to who controls the data and has the right to disseminate the data pools.

The issue of the increase in use of energy (electricity) to power this connected world has been a concern in view of global warming. In 2016, Swedish researcher, Anders Andrae, estimated that global computing power demand from interconnected devices, high-resolution video streaming, emails, surveillance cameras, and smart TVs could use 20% of global electricity by 2025. This draw on electricity is from 8.4 billion connected devices in 2017, and these devices will increase to 20.4 billion Internet of things by 2020. Such a large array of electronic devices means also that digital technologies are increasingly taking on complex command and control monitoring tasks along with systems that are fitted with advanced artificial intelligence that make their own decisions based on machine learning. What this means is that there is a worry that our society is increasingly dependent on interconnected structures that can be catastrophic if these systems fail or if they experience cyber attacks.

Labor economists have warned of the increase in unemployment should digitalization and artificial intelligence be adopted increasingly in our socio-economic lives and work routines (Robinson 2018). By the very nature of the characteristics of their business operations, the new IT companies such as Google and Microsoft also generate tremendous amount of profits with a smaller workforce. Google, for example, in 2012 had a profit of $14 billion operating with only a workforce of about 38,000 people whereas if one compares this to another industry such as the automotive manufacturers; General Motors, at its peak employment had 840,000 workers in its workforce earning a profit of $11 billion in 1979 (Ford 2015: 76). This means that the number of jobs in an IT operation by its very nature of what it produces will never be as high as in the traditional manufacturing sector such as the making of automobiles and appliances.

With the expansion of artificial intelligence-powered automation increasingly in all aspects of our lives, University of Oxford researchers in a study in 2017 has estimated an alarming figure of nearly half of all U.S. jobs will be at risk (Totty 2019). Such an alarming call is because with AI not only can it take over routine type work, it can also by its nature increasingly able to take over non-routine work through deep learning, use of Big Data, and so on. The business models of IT companies such as Facebook and Google do not need increasingly more workers but better algorithms. In view of this, routine and repetitive tasks jobs will be lost as AI systems using voice recognitions and natural language processing will replace them increasingly as these AI systems continue to improve.

In view of the above, it has been proclaimed by political leaders and labor and management economists that new forms of employment will be generated and the value of certain occupational groups will either be raised or lowered depending on the type of work performed. For example, jobs that require people skills such as kindergarten teachers will be harder to replace than college professors; radiologists can be replaced much easier than nurses. Those jobs that are less relevant will be eliminated and new ones added. The *Wall Street Journal* has reported that where robots have replaced humans in mundane jobs, new jobs such as robot minders are introduced to monitor and assist in directing the robots as they go through their assigned tasks (Mims 2019). With self-driven trucks/vehicles to be introduced in the future, it will require remote control to navigate these vehicles after delivering their consignments through suburban neighborhoods so that these immense size vehicles can get onto freeways. The truck driver will now be piloting the vehicle remotely just like the pilot of drones delivering their cargoes or weapons payload, instead of behind the wheel/joystick of the truck or the airplane.

As well, AI systems will also require humans as companions to train, explain, and sustain them. There will the need for thousands of human workers all over the world to train these systems and teach them how to operate cognitive tasks that we humans do naturally. These robotic AI systems will work alongside human workers in a variety of work environments.

This shift has always occurred when new technical innovations have been introduced in world history, and the digital revolution by no means is the only revolution that has and can generate this type of outcome. The increase in the use of AI could mean that efficiency and productivity will increase that might lead to shorter workdays thus allowing each worker more time for their other activities beyond work. If this is the case, it might mean that a better quality of lifestyle, and the issue of meaning that increasingly has been raised in our lifestyle reproduction can now be addressed and experienced. French workers, under a socialist government, have already enjoyed a 35-hour workweek regardless of whether the country has achieved the digitalization levels that has increased the country's productivity to warrant for such a length of a workweek.

With the robotization of labor, there will be less and less of a need for a medium-skilled workforce, and there has been the fear on the part of trade unions that worker's rights and wages will be systematically under threat. The grand bargain between workers, unions, and employers will increasingly be strained. As always, the liberal left will fight for social justice. The demand for a guaranteed basic wage scheme will be insisted in view that the normal period of a workday and workweek will change in light of the increased productivity.

As we know, with the advent of the Internet of things, the need and demand for workers skilled in engineering or sciences have been the modus operandi of the day. It has been argued by some that in this highly engineered Internet of things world, we need philosophers, experts in law, and political theorists who can bring to the table, skill sets and values that are beyond an algorithm, and who can anticipate potential violation of fundamental rights and privileges that we as a human society have shared over the last five thousand years of human history, and to fight for the rights of local communities to decide their own rules and laws.

The ubiquity of digital objects from laptops, Ipads, smart phones to digitalized "wearables" along with the associated apps available for us to interface means that we inform the corporations about what we in our work space and in our individual lives enjoy doing on our own time or have to do in our work or professional lives. In turn, the purveyors of the apps and the sellers of these digitalized equipment also have the power to shape our individual habits and wants. By expressing what we like and dislike online in responding to social media, we are being shaped by the app designers and corporations changing our values and beliefs

In this regard, among some users, the time spent on interacting in cyberspace can be extremely excessive. This has led to an irresistible urge to the screens that has made some feel they are losing their autonomy when it comes to directing their individual needs and wants (see, for example, Newport 2019). There is also the negative impact on the individual's psychological well-being. Some have reported on how social media can affect their moods. Compulsive behavior has resulted. As a result of this, there has also been a call to practicing digital minimalism (Newport 2019). Digital minimalism calls for us to adopt a philosophy toward this digital technology that we spent a limited time online on activities that we value, and hopefully this will miss everything that is not important to our lives. Whether such calls will be effective remains to be seen.

How then do we deal with these issues that have been raised in this chapter? In this unequal world whereby an estimated 2 billion people have no access to information and communication technologies, it will be even harder (WBGU 2019). With a burgeoning population, even though there are population losses during Dark Ages as I have indicated in the previous pages, and with diminishing natural resources, the survival of the planet is at stake. How do we meet the dangers of a changing environment facing us so that social and natural evolution can continue is a huge challenge that we all must keep in mind? The basis by which we make the wise choices is yet to be determined. For sure, AI will be pervasive in our everyday life practices along with robotic systems that will increasingly be present in the private and public spheres of our lives. Perhaps as James Lovelock (2019: 104–105) has recently projected in his book, *Novacene*, humankind will continue on in "an age in which humans and cyborgs would live together in peace—perhaps in loving grace—because they share a common project to ensure their survival. That project is maintaining the Earth as a livable planet."

Chapter 6

Reflections

If history is our teacher of life, what can we anticipate our future to be like? We know from historical events over the last five thousand years of world history there has been similar type of historical circumstances of such crisis periods—known as the Dark Ages—that we are experiencing now. However, despite these Dark Ages, social and natural evolution continued. Furthermore, we also know that these crises precipitated the emergence of new structures in the areas of technology, social and political organizations, and ideologies that led to further development of the human enterprise.

In light of the above, to me, our most important priority and strategic interest, at this point of time, is to ensure that social and natural evolution continue uninterrupted, albeit along paths that we as a human species, because of our excesses, have impacted on these evolutionary trajectories. To ensure that social and natural evolution progresses, we need to reconsider certain social factors and conditions existing that will continue to put tremendous stress on our planet and can prolong the crisis period if our natural environment is not allowed to recover.

Many scholars over the past thirty years have rung the emergency bells. Edward O. Wilson sounded the alarm in his books, *Half-Earth* and *The Future of Life*, of the dangers to our ecosystems, biodiversity and large-scale species extinction caused by human activities. Indicting humanity, Wilson joins the others in noting how we have decimated the natural environment and drawn down the nonrenewable resources of the planet. In doing so, according to Wilson, we have sped up the destruction of the ecosystems of many species and caused the extinction of some. The late U.S. environmentalist, David Brower, suggested the use of *CPR* to resuscitate our planet Earth. In this context, Brower (1995) suggested his *CPR* was *C* for conservation, *P* for preservation, and *R* for restoration of our natural environment. Both of these scholars have emphasized the natural environment's part of society's reproduction equation that has been impacted and consequently, the natural evolutionary process. Whereas other scholars, such as Paul and Anne Ehrlich (2004), have warned us of the dangers of overconsumption and population increases, and the threats they have on the social reproduction of the planet. The consumption and population increase factors need to be linked to the natural system limits discussion of available natural resources in light of exuberant lifestyles of the North. Even as early as the 1970s, there were already warnings from the work of the Club of Rome that the natural system has limits in providing resources for the reproduction of our social system (Meadows et al. 1972; Mesarovic and Pestel 1974). Exponential growth in population and the economy are key factors in accelerating the consumption of natural resources. Population growth and the increase in capital stock drives up the exponential curve of resource consumption. Even at that time, the Club of Rome was already warning of the measured pollution levels such as carbon dioxide and toxic wastes that were growing exponentially, and in the long run, through feedback loops, would also pose a limit to economic growth and generate disaster for our environment. In the second follow-up report, *Mankind at the Turning Point: The Second Report to the Club of Rome*, the global warning turned into an alert of various crises that the world faced. These crises covered the areas of population, environment, food, energy, and raw materials. In this second report, the Club of Rome was urging the world community to resolve these long-term crises at the global level. It is a long-term emergency to borrow the term from James Kunstler (2003). An update to the two reports to the Club of Rome was undertaken again by Donella Meadows and her colleagues in *Limits to Growth: The 30-Year Update* in 2004. In this follow-up, they again reiterated the warnings of natural resource scarcity and pollution issues that were degrading the landscape, and the threats posed to the future survivability of the planet.

Along this track, by no means was there a pause in these global warnings, especially coming from multilateral institutions and/or global committees headed by eminent state persons. Next to step up was the former prime minister of Norway, Gro Harlem Brundtland. As Chairperson of the World Commission on Environment and Development sponsored by the United Nations, a report, *Our Common Future*, was issued in 1987 echoing the same alert warnings made by the Club of Rome. The usual signs of environmental degradation in the areas of natural resource scarcity, species extinction (the consequence of habitat alteration and acidification of the environment), atmospheric pollution (greenhouse gases) leading to global warming (sounds familiar?), and deforestation were identified. World Commissions, such as those headed by Gro Brundtland and Willy Brandt (1987), were not the only institutions joining the Club of Rome's emergency calls. The United Nations further chimed in. Five years after the Brundtland Commission reported its warnings, a global conference on the environment was organized under the auspices of the United Nations in Rio de Janeiro in 1992. Named the United Nations Conference on Environment and Development, commonly known as the Earth Summit, this global meeting identified the natural resource limits and scarcities the global community faced, and the causes and consequences of such tendencies. The identified natural resource scarcities and outcomes, the pollution levels, and consequences as a result of uncontrolled human economic and population growth announced and discussed at the summit were by no means new insights to what the global community's elites and politicians were not already aware of or on their part chose not to hear. Simply put, in UN fashion, it was repeated again that a global agreement was needed to handle these natural resource and pollution crises that the world community faced. With the impending shrinking availability of natural resources and global warming that will impact on socioeconomic life, the solution that came out of the summit was a call for *sustainable development* as the theme that all UN participating nation-states should adopt as their national programs and policies. Of course, the United Nations did not intend to be left out of the overall coordinating role for this proclamation of sustainable development, as this was its preserve and its *raison d'être*. Like the conclusion of every UN summit, a customary set of policy guidelines was voted on and established which contained twenty-one principles for action—Agenda 21—that the members of the UN were supposed to follow.

With the above set of outcomes, we are off to the races in terms of undertaking scientific studies through grants, coordination of multilateral and bilateral development programs, regional and global coordinating meetings and conferences, expert visits to projects in the South, expert multilateral and bilateral donor mission team meetings to coordinate regional, national, and global efforts. Other multilateral global agencies joined in and were also involved in not only providing their expertise, but as well, their financial resource allocations for the regions in the South. One of the major players in this donor coordinating and funding exercise was the International Bank for Reconstruction and Development (IBRD), commonly known as the World Bank and its World Bank Group affiliates. Notwithstanding this, major UN multilateral agencies were also involved—such as United Nations Development Program (UNDP), United Nations Children Fund (UNICEF), and the United Nations Environment Program (UNEP), and the World Health Organization (WHO)—along with the respective bilateral foreign aid development agencies (such as U.S. Aid) of the developed nations.

At this late timing of the game, it is important to note that for us to traverse this current global crisis we are experiencing, those factors (natural resource and energy scarcities, global population growth, overconsumption, habitat destruction, global air pollution, and climate warming) that were identified over the course of the last thirty or so years need to be kept in the overall equation for us to decide which strategic options we take in order for a smooth system transition to proceed, and that social and natural evolution continue uninterrupted. There is too much emphasis these days just on climate change and global warming as if this was the only major threat to the planet. As we have been told by numerous studies that climate change cannot be solved just by reducing greenhouse gases alone; an interconnected set of loops and feedbacks need to be considered. In this context, Lovelock's (1979) Gaia Hypothesis should be remembered. All aspects of our social reproduction from consumption behavior, procreation, lifestyles, and so on have to be examined as they all have an impact on the survivability of our planet. The development of deep learning machines and AI can assist in this regard.

On an overall basis, two factors, namely population growth and overconsumption, have still to be highlighted and weighted to the highest order in the planet's survivability equation unless different socioeconomic and ideological structures emerge—such as the bioregional and degrowth models—and they become the new organizing principles of our societal systems. For the latter to occur as angels of history, it will mean that the Dark Age conditions have reached such a stage that the existing socioeconomic and political institutions and ideologies we have in place now no longer can provide the capacity to reproduce the system in the long run, and system transition follows. To date, this is not too likely in the medium term. The global crises have not developed to this stage yet. Take, for example, the issue of diminishing supply of energy resources, such as fossil fuel. The issue of availability has been modulated and tempered down by differences in living standards within and between the North and the South. If we narrow it down to the fuel consumption of two global hegemonic countries such as the United States and the PRC, just in terms of fossil fuel usage, the United States has less than 5% of the world's population but uses a quarter of the global fossil fuel (Worldwatch 2018). If we add to this, the PRC usage, though it is not at the same developmental status as the United States yet, the combined consumption levels of fossil fuels of these two nations would impact on the availability of this natural resource, especially when we have gone past the Peak Oil level (Chew 2008). But it seems that differences within each nation, i.e. unequal energy consumption of different classes in the United States and in the PRC, and differences between nations, have managed to keep the fossil fuel levels flowing, albeit these fuel supplies have been reinforced and protected by military forces and diplomatic/trade initiatives, and also advances in oil extracting technologies (shale oil) via digitalization and AI as well. Global disease pandemics can also be a significant factor in reducing energy consumption in the area of fossil fuels, and as we know, the appearance of disease pandemics is also a feature of Dark Ages as we have stated in the previous chapters.

In the medium term, what we need is to focus and strategize so that natural and social evolution continue unabated. To do this, I have offered some contemplations in the previous chapters to address the current Dark Age conditions of economic slowdowns, political unrest and breakdowns, deurbanization, increased migration, biodiversity crisis, landscape transformations, disease pandemics, and climate changes. Whether the technological advances and the Internet of things should continue to be adopted to address these natural and social ecological crises conditions is no longer an issue, other than an academic debate. It should not matter much now because they are already part of our lives, and how we organize our daily activities. Most of us are already mesmerized by these advances even the luddites and the intellectual elites. What we need to consider is to ensure that these technological advances and things are utilized in such a way that will take into consideration the limits of our natural environment and its resources so that natural and social evolution will continue.

In this era of global crisis or digital Dark Age, what has been the response to the technological advances in the area of robotics, AI, machine learning, and virtual/augmented realities? I have discussed some of the main reactions to these innovative ideas in the previous chapter. To me, most of the reactions follow in a standard bureaucratic manner: the need to establish rules and regulations to control and direct the use of these technological advances and Internet of things so that civil and criminal actions do not take place, and our privacy is not violated. The underlying protocol is to protect humanity for humanity's sake.

What needs to be contemplated and meditated is the framing on how we conduct and rationalize our actions and laws from now on in this "new" world of robotics and AI. We should break away from the anthropocentric iron cage now gilded with human values, human rights, human ethics that most, if not all, of us used as a basis to rationalize our actions. This manner of continuing to determine our actions in the upcoming "new" world is still very much prevalent in most of the recent writings on the advent of AI in our lives (see, for example, Coleman 2019). Instead, we need to re-conceptualize and consider new values, rights, and ethics that *are not human centered or anthropocentric in nature* when it comes to the new world of robotics and AI that we are immersing into. As a basic starting point, we should award intrinsic rights and value to humans, other living things, and AI developed systems, as the basic principle or charter underlying our social, economic, and political organization in our future systems. In this way, no entity has priority, and that everyone including AI systems has the right to evolve as well, without controls imposed by humans on these AI systems. It will be another step in the overcoming of the limitation of our human mind. This will dovetail with the recent proposal of James Lovelock in his latest book *Novascene*, that in the end, AI systems and us will need to coexist in peaceful harmony. The aim will be to ensure that human life and other AI-based systems continue to co-evolve on Planet Earth in a sustainable and livable manner. In the long term, the system transition and expansion that are expected to occur following the end of this digital Dark Age will have the human species in collaboration with AI systems and cyborgs pursuing space explorations, and perhaps, even off-Earth living. This will be a new stage of social and natural evolution. In this light, world history continues!

Bibliography

Aberley, Doug. 1999. "Interpreting Bioregionalism." In *Bioregionalism*, edited by Michael McGinnis, 13–42. London: Routledge.

Adam, Barbara. 2004. *Time*. Cambridge, MA: Polity Press.

Adams, Madonna. 1997. "Environmental Ethics in Plato's Timaeus." In *Environment and the Greeks*, edited by L. Westra and T. Robinson. London: Rowman and Littlefield.

Alexander, Samuel, and Gleeson, Brenden. 2019. *Degrowth in the Suburbs*. London: Palgrave-MacMillan.

Alexandratos, N., and Bruinsma, J. 2012. *World Agriculture: Towards 2030/50*. Rome: FAO.

Alisa, Giacomo, Demaria, F., and Kallis, Giorgos (eds). 2015. *Degrowth*. London: Routledge.

Amin, Samir, Arrighi, Giovanni, Frank, Andre Gunder, and Wallerstein, Immanuel M. 1982. *Dynamics of Global Crisis*. New York: Monthly Review Press.

Anthanasius. 1980. *The Life of Antony and the Letter to Marcellinus*. Translated by R. C. Griegs. New York: Paulist Press.

Bailenson, Jeremy. 2018. *Experience on Demand: What Virtual Reality Is, How It Works and What It Can Do*. New York: Norton.

Baldwin, Richard. 2019. *The Globotics Upheaval*. Oxford: Oxford University Press.

Barrat, James. 2015. *Our Final Invention*. New York: Thomas Dunne.

Baudrillard, Jean. 1975. *For a Critique of the Political Economy of the Sign*. St. Louis: Telos Press.

———. 1983. *Simulations*. Paris: Semiotexte.

———. 1994. *The Illusion of the End*. Stanford: Stanford University Press.

———. 1998. *The Consumer Society*. Beverly Hills: Sage Publications.

Beaujard, Philippe. 2005. "The Indian Ocean in Eurasian and African World-Systems Before the Sixteenth Century." *Journal of World History*, Vol. 16, No. 4: 411–465.

———. 2009. "Evolution and Temporal Delimitations of Possible Bronze Age World-Systems in Western Asia, Africa and the Mediterranean." In *Interweaving*

Worlds: Systemic Interactions in Eurasia 7th to 1st Millennia BC, edited by T. Wilkinson, et al. Oxford: Oxbow Press.

———. 2010. "From Three Possible Iron-Age World-Systems to a Single Afro-Eurasian World-System." *Journal of World History*, Vol. 21, No. 1: 1–43.

Berg, Peter. 1978. *Reinhabiting a Separate Country*. San Francisco, CA: Planet Drum Foundation.

———. 2009. *Envisioning Sustainability*. San Francisco: Subculture Books.

Berg, Peter, and Dasmann, Ray. 1978. "Reinhabiting California." In *Reinhabiting a Separate Country*, edited by Peter Berg, 217–220. San Francisco, CA: Planet Drum Foundation.

Bergmann, Werner. 1985. "Das Frühe Mönchtum als soziale Bewegung." *Kölner Zeitschrift für Soziologie und Sozialpsychologie*, Vol. 37, No. 1: 30–59.

Boulding, Kenneth E. 1966. "The Economics of the Coming Spaceship Earth." In *Environmental Quality in a Growing Economy*, edited by Henry Jarrett, 14–25. Baltimore: Johns Hopkins University Press.

Bramble, Portia Boni. 2009. *A Survivor's Guide to Running the Farmer's Market.* M.A. Thesis. Humboldt State University, Arcata, California.

Brandt Commission. 1987. *Common Crisis*. London: Pan.

Braudel, Fernand. 1972. *The Mediterranean and the Mediterranean World in the Age of Philip II.* Vol. 1. London: Fontana.

———. 1981. *The Structure of Everyday Life*. Vol. 1. New York: Harper and Row.

———. 1982. *The Wheels of Commerce*. Vol. 2. New York: Harper and Row.

———. 1984. *The Perspective of the World.* Vol. 3. New York: Harper and Row.

———. 1989. *The Identity of France*. Vols. 1–2. New York: Harper and Row.

———. 2001. *Memory and the Mediterranean.* New York: Alfred Knopf.

Bronson, Kelly, and Knezevic, Irena. 2016. "Big Data in Food and Agriculture." *Big Data and Society*, January–June: 1–5.

Brower, David. 1995. "CPR for the Earth." *Humboldt Journal of Social Relations*, Vol. 21, No. 1: 1–10.

Brown, Lester R. 2006. *Plan B2.0.* New York: Norton.

———. 2009. *Plan B4.0.* New York: Norton.

———. 2015. *The Great Transition*. New York: Norton.

Brown, Peter. 1978. *The Making of Late Antiquity*. Cambridge, MA: Harvard University Press.

———. 1989. *The World of Late Antiquity*. New York: Norton.

Brundtland, Gro Harlem. 1987. *Our Common Future*. New York: Oxford University Press.

Brynjolfsson, Erik, and McAfee, Andrew. 2014. *The Second Machine Age*. New York: Norton.

Buch-Hansen, Hubert, and Koch, Max. 2019. "Degrowth Through Income and Wealth CAPS." *Ecological Economics*, Vol. 160, June: 264–271.

Burroughs, William J. 2005. *Climate Change in Prehistory*. Cambridge, MA: Cambridge University Press.

Caldwell, Melissa. 2018. "Hacking the Food System." In *Digital Food Activism*, edited by Tanja Schneider, et al. London: Routledge.

Campbell, C. 2004. *The Coming Oil Crisis*. Essex: Multiscience Publishing Co.
Cantor, Norman F. 1994. *The Civilization of the Middle Ages*. New York: Harper.
Carr, Mike. 2005. *Bioregionalism and Civil Society: Democratic Challenges to Corporate Globalism*. Vancouver: University of British Columbia Press.
Cartwright, F. E. 1972. *Disease and History*. New York: Crowell.
Castellanos, Sara. 2017. "Volkswagen to Use Deep Learning to Optimize Traffic Flows." *Wall Street Journal*, June 28, 2017.
Castells, Manuel. 1989. *The Informational City*. Oxford: Oxford University Press.
———. 1996. *The Rise of Network Society*. Oxford: Blackwell.
———. 1997. *The Power of Identity*. Boston: Blackwell.
———. 1998. *The End of the Millennium*. Boston: Blackwell.
———. 2000. *The Network Soci*ety. 2nd Edition. Boston: Blackwell.
Castells, Manuel, Fernandez-Ardevol, M., Qiu, J. L., and Sey, A. 2007. *Mobile Communication and Society*. Cambridge, MA: MIT Press.
Castells, Manuel, and Himanen, Pekka. 2002. *The Information Society and the Welfare State: The Finnish Model*. Oxford: Oxford University Press.
Castles, Stephen, and Miller, Mark J. 2003. *The Age of Migration*. 2nd Edition. New York: Guilford Press.
Center for Responsible Travel. 2018. *The Case for Responsible Travel: Trends and Statistics*. Washington, DC: CRT.
Chase-Dunn, C., and Anderson, E. N. 2005. *The Historical Evolution of the World System*. London: Palgrave.
Chase-Dunn, C., and Hall, T. 1997. *Rise and Demise: Comparing World-Systems*. Boulder: Westview Press.
Chernykh, E. N. 1992. *Ancient Metallurgy in the USSR: The Early Metal Age*. Cambridge, MA: Cambridge University Press.
Chew, Sing C. 1997. "For Nature: Deep Greening World-Systems Analysis for the 21st Century." *Journal of World-Systems Research*, Vol. 3, No. 3: 381–402.
———. 1999. "Ecological Relations and the Decline of Civilizations in the Bronze Age World System: Mesopotamia and Harrapa 2500 B.C.–1700 B.C." In *Ecology and the World System*, edited by W. Goldfrank, et al. Greenwich, CT: Greenwood Press.
———. 2001. *World Ecological Degradation: Accumulation, Urbanization, and Deforestation 3000BC–AD2000*. Lanham, MD: AltaMira Press/Rowman and Littlefield Publishers.
———. 2002. "Globalization, Dark Ages, and Ecological Degradation" *Global Society*, Vol. 16, No. 4, October: 333–356.
———. 2003. "Ecology in Command." In *Structure, Culture, and History: Recent Issues in Social Theory*, edited by Sing C. Chew and David Knottnerus. Lanham, MD: Rowman and Littlefield Publishers.
———. 2005. "From Harappa to Mesopotamia and Egypt to Mycenae: Dark Ages, Hegemonial Shifts, and Environmental/Climate Changes." In *The Historical Evolution of the World Systems*, edited by Christopher Chase-Dunn and E. N. Anderson. London: Palgrave.

———. 2006. *The Recurring Dark Ages: Ecological Stress, Climate Changes, and System Transformation*. Lanham, MD: AltaMira Press/Rowman and Littlefield Publishers.

———. 2007. *Ecological Futures: What History Can Teach Us*. Lanham, MD: AltaMira Press/Rowman and Littlefield Publishers.

———. 2016. "Nature-Culture Relations: Early Globalization, Climate Changes, and System Crisis." *Sustainability*, Vol. 8, No. 78. DOI: 10.3390/su8010078.

———. 2017. "Southeast Asia in the First Eurasian World Economy." In *Comparing Globalization: Historical and World-System Approaches*, edited by Thomas D. Hall. New York/Berlin: Springer.

———. 2018. *The Southeast Asia Connection: Trade and Polities in the Eurasian World Economy 500BC–AD500*. New York: Berghahn Press.

———. 2020. "Introduction." In *Bill Devall Living Deep Ecology: A Bioregional Journey*, 3–10. Lanham, MD: Rowman and Littlefield.

Childe, V. Gordon. 1942. *What Happened in History*. London: Penguin.

Chirot, Daniel, and Montgomery, Scott. 2015. *The Shape of the New: Four Big Ideas and How They Made the Modern World*. Princeton, NJ: Princeton University Press.

Chui, Michael, Manyika, J., Miremodi, M., Henke, N., Chung, Rita, Nel, Pieter, and Malhotra, S. 2018. *Notes from the AI Frontier: Insights from Hundred Use Cases*. Discussion Paper. San Francisco, CA: Mckinsey and Co.

Cicero, Marcus Tullius. 1942. *On The Orator: Books 1–2 Rhetorical Treatises*. Cambridge, MA: Harvard University Press.

Constanza, Robert, Grumlich, Lisa, and Steffen, Will. 2007. *Sustainability or Collapse: An Integrated History and Future of People on Earth*. Boston: MIT Press.

Daly, Herman (ed.). 1973. *Toward a Steady State Economy*. San Francisco: W.H. Freeman.

———. 1996. *Beyond Growth*. Boston: Beacon Press.

Daniel-Rops, H. 2001. *The Church in the Dark Ages*. London: Phoenix Press.

Dauvergne, Peter. 2020a. "The Globalization of Artificial Intelligence: Consequences for the Politics of Environmentalism." *Globalizations* 1–17.
DO1: 10.1080/14747731.2020.

———. 2020b. *AI in the Wild: Sustainability in the Age of Artificial Intelligence*. Boston: MIT Press.

Davenport, Thomas H. 2018. *The AI Advantage*. Cambridge, MA: MIT Press.

Deffeyes, K. S. 2001. *Hubbert's Peak: The Impending World Oil Shortage*. Princeton, NJ: Princeton University Press.

DeLello, Courtney. 2004. *Community Supported Agriculture: A Case Study of Humboldt County*. M.A. Thesis. Humboldt State University, Arcata, California.

Denemark, R., et al. 2000. *World System History*. London: Routledge.

Desborough, V. R. 1972. *The Greek Dark Ages*. London: Ernest Benn.

Devall, Bill. 1988. *Simple in Means Rich in End*s. Salt Lake City, UT: Gibbs Smith.

———. 1993. *Living Richly in an Age of Limits*. Salt Lake City, UT: Gibbs Smith.

———. 2020. *Living Deep Ecology: A Bioregional Journey*. Lanham, MD: Rowman and Littlefield.

Dodds, E. R. 1965. *Pagan and Christian in an Age of Anxiety: Some Aspects of Religious Experience from Marcus Aurelius to Constantine*. Cambridge, MA: Cambridge University Press.

Dodge, Jim. 1981. "Living by Life: Some Bioregional Theory and Practice." *CoEvolution Quarterly*, Vol. 32: 6–12.

Domingos, Pedro. 2015. *The Master Algorithm*. New York: Basic Books.

Drake, Brandon L. 2012. "The Influence of Climate Change on the Late Bronze Age Collapse and the Greek Dark Ages." *Journal of Archaeological Science*. DOI: 10.1016/j.jas.2012.01.029.

Drews, R. 1993. *The End of the Bronze* Age. Princeton, NJ: Princeton University Press.

Duckett, Eleanor S. 1990. *Gateway to the Middle Ages: Monasticism*. London: Dorset Press.

Dunlap, Riley (ed.). 2002. *Social Theory and the Environment*. Lanham, MD: Rowman and Littlefield.

Dunn, Marilyn. 2000. *The Emergence of Monasticism: From Desert Fathers to the Early Middle Ages*. Oxford: Blackwell.

Dyck, Bruno. 1992. *Inside the Food Systems: How do Community Supported Farm Work?* Marketing Digest.

Eckhardt, William. 1992. *Civilizations, Empires and Wars: A Quantitative History of War*. Jefferson, NC: McFarland.

Eder, Klaus. 1996. *The Social Construction of Nature*. Thousand Oaks, CA: Sage Publications.

Ehrlich, Paul, and Ehrlich, Anne. 2004. *One with Nineveh: Politics, Consumption, and the Human Future*. Washington, DC: Island Press.

Ford, Martin. 2015. *The Rise of the Robots*. New York: Basic Books.

———. 2018. *Architects of Intelligence: The Truth About AI from the People Building It*. Birmingham, UK: Packt Publishing.

Foster, John Bellamy. 1999a. "Marx's Theory of Metabolic Rift: Classical Foundations for Environmental Sociology." *American Journal of Sociology*, Vol. 105: 366–405.

———. 1999b. *The Vulnerable Planet: A Short Economic History of the Environment*. New York: Monthly Review Press.

———. 2000. *Marx's Ecology: Materialism and Nature*. New York: Monthly Review Press.

Frank, Andre Gunder. 1991. "Transitional Ideological Modes: Feudalism, Capitalism, and Socialism." *Critique of Anthropology*, Vol. 11, No. 2: 171–188.

———. 1993a. *World System 500 Years or 5,000 Years*. London: Routledge.

———. 1993b. "Bronze Age World System Cycles." *Current Anthropology*, Vol. 34, No. 4, October: 383–429.

———. 1998. *ReOrient: Global Economy in the Asian Age*. Berkeley: University of California Press.

———. 2015. *Reorienting the 19th Century*. London: Routledge.

Frank, Andre Gunder, and Gills, Barry. 1992. "The Five Thousand Year Old System: An Interdisciplinary Introduction." *Humboldt Journal of Social Relations*, Vol. 18, No. 1: 1–55.

Franklin, C. V., Havener, I., and Francis, J. (Trans.). 1981. *Early Monastic Rules: The Rules of the Fathers and the Regula Orientalis*. Collegeville: The Liturgical Press.

Frend, W. H. C. 1972. "The Monks and the Survival of the Eastern Roman Empire in the 5th Century." *Past and Present*, Vol. 54: 3–24.

———. 1984. *The Rise of Christianity*. Philadelphia: Fortress Press.

Friedman, Jonathan. 2004. "Globalization, Transnationalization and Migration." In *Worlds on the Move*, edited by J. Friedman and Shalima Randeira, 63–90. London: I. B. Tauris.

Friedman, Jonathan, and Chase-Dunn, Chris. 2016. *Hegemonic Declines: Past and Present*. London: Routledge.

Friedrichs, Jorg. 2013. *The Future is Not What It Used to Be*. Boston: MIT Press.

German Advisory Council on Global Change. 2019. *Towards Our Common Digital Future*. Berlin: WBGU Secretariat.

Gills, Barry K., and Thompson, William R. 2006. *Globalization and Global History*. London: Routledge.

Glacken, Clarence. 1967. *Traces of the Rhodian Shore*. Berkeley, CA: University of California Press.

Goehring, J. E. 1990. "The World Engaged: The Social and Economic World of Early Egyptian Monasticism." In *Gnosticism and the Early Christian World*, edited by J. E. Goehring, C. H. Hedrick, and J. T. Sanders, 34–42. Sonoma: Polebridge Press.

Goldin, Owen. 1997. "The Ecology of Critias and Platonic Metaphysics." In *Environment and the Greeks*, edited by L. Westra and T. Robinson. London: Rowman and Littlefield.

Gorz, André. 1982. *Farewell to the Working Class: An Essay on Post Industrial Society*. Boston: South End Press.

Gould, Kenneth A., Pellow, David N., and Schnaiberg, Allan. 2016. *The Treadmill of Production*. New York: Routledge.

Green, Henry A. 1986. "The Socioeconomic Background of Christianity in Egypt." In *The Roots of Egyptian Christianity*, edited by Birger A. Pearson and James Goehring, 100–113. Philadelphia, PA: Fortress Press.

Griggs, C. W. 1990. *Early Egyptian Christianity from its Origins to 451 C.E.* Leiden: Kluwer.

Habermas, Jürgen. 1981. "Modernity Versus Postmodernity." *New German Critique*, No. 22, Winter: 3–14.

———. 1987. *The Philosophical Discourse of Modernity*. Cambridge, MA: MIT Press.

Hansson, Carina Borgstrom. 2003. "Misplaced Concreteness and Concrete Places." In *Lund Studies in Human Ecology*. Vol. 7. Lund: Lund University.

Hardy, Edward Rochie. 1952. *Christian Egypt: Church and People*. New York: Oxford University Press.

Harrison, Robert Pogue. 1992. *The Shadow of Civilization*. Chicago: University of Chicago Press.

Harvey, David. 1990. *The Condition of Postmodernity*. Oxford: Blackwell.

———. 2003. *New Imperialism*. Oxford: Oxford University Press.

———. 2018. *Marx, Capital and the Madness of Economic Reason*. Oxford: Oxford University Press.

Hawken, Paul. 1999. *Natural Capitalism: Creating the Next Industrial Revolution*. Boston: Little Brown and Co.

Heather, Peter. 1991. *Goths and Romans*. Oxford: Clarendon Press.

———. 1996. *The Goths*. Oxford: Blackwell.

———. 2006. *The Fall of the Roman Empire: A New History of Rome and the Barbarians*. Oxford: Oxford University Press.

Heather, Peter, and Matthews, John. 1991. *The Goths in the Fourth Century*. Liverpool: Liverpool University Press.

Henderson, E., and Van Eyn, Robyn. 1999. *Sharing the Harvest: A Guide to CSA*. White River Junction, VT: Chelsea Green.

Herrera, Sebastian. 2019. "How Amazon's Empire is Challenging UPS and Fedex." *Wall Street Journal*, August 29, 2019.

Hodges, Richard, and Whitehouse, David. 1983. *Mohammed, Charlemagne and the Origins of Europe*. Ithaca: Cornell University Press.

Holslin, Peter. 2015. "Virtual Reality Slaughterhouses Could Be the Future of Animal Rights Activism." *Vice*. https://www.vice.com/enus/article/4wb5km/virtual-reality-is-the-new-frontier-in-animal-activism-511.

Homer-Dixon, T., and Blitt, J. 1998. *Ecoviolence–Links Among Environment, Population and Security*. Lanham, MD: Rowman and Littlefield.

Hopkins, T., and Wallerstein, I. 1996. *The Age of Transition: Trajectory of the World System*. Atlantic Heights, NJ: Pluto.

———. 1998. *Utopistics*. New York: The New Press.

House, Freeman. 1998. *Totem Salmon*. Boston: Beacon Press.

Hughes, J. Donald. 1975. *Ecology in Ancient Civilizations*. Albuquerque: University of New Mexico Press.

———. 1994. *Pan's Travail*. Baltimore: Johns Hopkins University Press.

———. 2006. *What is Environmental History?* Cambridge, MA: Polity Press.

Huntington, Samuel D. 1996. *The Clash of Civilizations*. New York: Simon and Schuster.

Husain, Amir. 2017. *The Sentient Machine*. New York: Scribner.

Innes, Matthew. 2000. *State and Society in the Early Middle Ages: The Middle Rhine Valley 400–1000*. Cambridge, MA: Cambridge University Press.

Intergovernmental Panel on Climate Change. 2001. *Impacts, Adaptation, and Vulnerabilities*. Cambridge, MA: Cambridge University Press.

International Tourism Society. 2000. *Ecotourism Statistical Fact Sheet*. Washington, DC.

James, Jeffrey. 2003. *Bridging the Global Digital Divide*. Cheltenham: Edward Elgar.

Jones, A. H. M. 1959. "Overtaxation and Decline of the Roman Empire." *Antiquity*, Vol. XXXIII: 235–250.

———. 1964. *The Later Roman Empire AD 244–602*. Oxford: Blackwell.

———. 1974. *The Roman Economy*. Totowa: Roman and Littlefield.

Kagami, Mitsuhiro, et al. (eds). 2004. *Information Technology Policy and the Digital Divide*. Cheltenham: Edward Elgar.

Kerschner, Christian, et al. (eds). 2018. "Technology and Degrowth." *Journal of Cleaner Production*, Vol. 197, Part 2, October.

Kimbrell, Andrew (ed.). 2002. *Fatal Harvest: The Tragedy of Industrial Agriculture.* Covelo, CA: Island Press.

Kristiansen, Kristian. 1993. *The Emergence of the European World System in the Bronze Age: Divergence, Convergence, and Social Evolution during the First and Second Millennia B.C. in Europe*. Sheffield Archaeological Monographs no. 6.

———. 1998. *Europe before History*. Cambridge, MA: Cambridge University Press.

Kristiansen, Kristian, and Larsson, Thomas. 2005. *The Rise of Bronze Age Society Travels, Transmission, and Transformation.* Cambridge, MA: Cambridge University Press.

———. 2007. "Household Economy, Long-Term Change, and Social Transformation: The Bronze Age Political Economy of Northwestern Europe." *Nature and Culture*, Vol. 2, No. 2, Autumn: 71–86.

Kuntsler, James Howard. 2005. *The Long Emergency*. New York: Grove Press.

Lawrence, C. H. 1984. *Medieval Monasticism: Forms of Religious Life in Western Europe During the Middle Ages*. London: Longman.

Leonardi, Emanuele. 2019. "Bringing Class Analysis Back In: Assessing the Transformation of the Value-Nature Nexus to Strengthen the Connection Between Degrowth and Environmental Justice." *Ecological Economics*, Vol. 156, February: 83–90.

Leopold, Aldo. 1949. *A Sand County Almanac*. New York: Oxford University Press.

Locke, John. 1891. *The Philosophy of Locke*. New York: Henry Holt and Company.

Lovelock, James. 1979. *Gaia: A New Look at Life on Earth*. Oxford: Oxford University Press.

———. 2019. *Novascene*. New York: Penguin.

Malone, E. E. 1950. *The Monk and the Martyr*. Washington, DC: Catholic University Press.

Manyika, James, Bugin, Jacques, Chui, Michael, and Silberg, Jake. 2018. *The Promise and Challenge of the Age of Artificial Intelligence*. San Francisco: McKinsey Global Institute.

Markantonatou, Maria. 2013. From the Limits to Growth to "Degrowth': Discourses of Critique of Growth in the Crises of the 1970s and 2008. *Working Paper 05/2013 der DFG-Kollegforscher lnnengruppe Postwachstumsgesellschaften*. Friedrich-Schiller-Universität Jena.

Marx, Karl. 1959. *Capital Vols. 1–3*. Moscow: Progress Publishers.

———. 1972. *Grundrisee*. London: Penguin.

McAfee, Andrew. 2018. "How the I-Phone Helped Saved the Planet." *Wired*, June 28, 2018.

———. 2019. *More from Less*. New York: Scribners.

McCormick, John. 2020. "California Utilities Hope Drones, AI Will Lower Risk of Future Wildfires." *Wall Street Journal*, September 11.

McCormick, M. 2001. *Origins of the European Economy*. Cambridge, MA: Cambridge University Press.

———. 2002. "New Light on the 'Dark Ages" How the Slave Trade Fuelled the Carolingian Economy." *Past and Present*, No. 177: 16–54.

McGinnis, M. (ed.). 1998. *Bioregionalism*. London: Routledge.

McKibben, Bill. 2007. *Deep Economy: The Wealth of Communities and the Durable Future*. New York: Times Books.

———. 2019. *Falter*. New York: Henry Holt.

McNeill, William H. 1975. *Plagues and History*. New York: Anchor Books.

———. 1984. "Human Migration in World Perspective." *Population and Development Review*, Vol. 10: 1–18.

McNeill, William H., and Adam, Ruth S. (eds). 1978. *Human Migration: Patterns and Policies*. Bloomington: Indiana University Press.

Meadows, Donella, et al. 1972. *The Limits to Growth*. New York: Signet.

———. 2004. *Limits to Growth: The 30th Year Update*. White River Junction, VT: Chelsea Green.

Mesarovic, Mihalo, and Pestel, Eduard. 1974. *Mankind at the Turning Point*. New York: Dutton.

Meyer, Robert T. (trans.). 1950. *St Athanasius the Life of Saint Antony*. New York: Newman Press.

Miles, Elizabeth. 2011. *From Farm to the Plate: Discourse and Perceptions of Local Food in Humboldt County*. M.A. Thesis. Humboldt State University, Arcata, California.

Millennium Ecosystem Assessment. 2003. *Ecosystems and Human Well-Being: A Framework for Assessment*. Covelo, CA: Island Press.

———. 2005. *Ecosystems and Human Well-Being*. Washington, DC: World Resources Institute.

Mims, Christopher. 2019. "The Next Hot Job: Pretending to be a Robot." *Wall Street Journal*, September 1, 2019.

Mitchell, William J. 1999. *ETopia*. Cambridge, MA: MIT Press.

Modelski, George. 2006. "Ages of Reorganization: Self-Organization in the World System." *Nature and Culture*, Vol. 1, No. 2: 205–227.

Modelski, George, and Thompson, William. 1999. "The Evolutionary Pulse of the World System: Hinterland Incursions and Migrations, 4000 BC to AD 1500." In *World System Theory in Practice*, edited by Nick Kardulias, 241–274. Lanham, MD: Rowman and Littlefield.

Mol, Arthur P. J. 2003. *Globalization and Environmental Reform*. Cambridge, MA: MIT Press.

Mol, Arthur P. J., and Sonnenfeld, David A. 2000. "Ecological Modernization Around the World Perspectives and Critical Debates." *Environmental Politics*, Vol. 9, No. 1: 1–14.

Moore, Jason W. 2016a. *Capitalism and the Web of Life*. New York: Verso.

——— (ed.). 2016b. *Anthropocene or Capitaloscene: Nature, History and the Crisis of Capitalism*. Oakland: PM Press.

Morris, Ian. 2013. *The Measure of Civilization: How Social Development Decides the Fate of Nations*. Princeton, NJ: Princeton University Press.

Muir, John. 1911. *The Mountains of California*. New York: The Century Co.

———. 1918. *A Thousand Mile Walk to the Gulf.* Boston: Houghton and Mifflin and Co.

———. 1986. *The Yosemite.* Madison: University of Wisconsin Press.

Muraca, Barbara. 2013. "*Décroissance*: A Project for a Radical Transformation of Society." *Environmental Values*, No. 22: 147–169.

Murphy, Cullen. 2007. *Are We Rome?* New York: Houghton Mifflin.

Nafeez, Mosaddeq Ahmed. 2010. *A User's Guide to the Crisis of Civilization.* London: Pluto Press.

———. 2017. *Failing States Collapsing Systems.* London: Springer.

Nelson, Anitra, and Schneider, Francois (eds). 2018. *Housing for Degrowth.* London: Routledge.

Newport, Cal. 2019. *Digital Minimalism.* New York: Portfolio.

O'Connor, James. 1988. "Capitalism, Nature, Socialism: A Theoretical Introduction."

Capitalism, Nature, Socialism, Vol. 1: 11–38.

———. 1998. "The Conditions of Production and the Production of

Conditions." In *Natural Causes: Essays in Ecological Marxism*, 135–178.

New York: Guilford.

Olson, Robert, and Rejeski, David. 2005. *Environmentalism and Technology of Tomorrow.* Washington, DC: Island Press.

Ostwalt, Philipp, and Rieniets, T. 2006. *Atlas of Shrinking Cities.* Ostfildem: Hatze Cantz Verlag.

Patel, Raj, and Moore, Jason. 2017. *History of the World in Seven Cheap Things.* Berkeley, CA: University of California Press.

Pearson, Birger, and Goehring, James (eds). 1986. *The Roots of Egyptian Christianity.* Philadelphia: Fortress Press.

Perlin, John. 1989. *A Forest Journey.* Cambridge, MA: Harvard University Press.

Pirenne, Henri. 1992. *Mohamed and Charlemagne.* New York: Barnes and Noble.

Price, David. 2020. "An AI Breaks the Writing Barrier." *Wall Street Journal*, August 22.

Roberts, Neil. 1998. *The Holocene.* Oxford: Blackwell.

Robinson, William I. 2014. *Global Capitalism and the Crisis of Humanity.* Cambridge, MA: Cambridge University Press.

———. 2018. "The Next Economic Crisis: Digital Capitalism and Global Police State." *Race and Class*, 1–16. DOI: 10.1177/0306396818769016journals.sagepub .come/home/rac.

Rousseau, S. 2012. *Food and Social Media: You are What You Tweet.* Lanham, MD: AltaMira Press.

Rubenson, S. 1995. *The Letters of St. Anthony Monasticism and the Making of a Saint.* Minneapolis: Fortress Press.

Rubin, Peter. 2018. *Future Presence: How Virtual Reality Is Changing Human Connection, Intimacy, and Limits of Ordinary Life.* New York: Harper.

Ruddiman, W. F., and Thomson, J. S. 2001. "The Case for Human Causes of Increased Atmospheric Methane over Last 5,000 Years." *Quaternary Science Review*, Vol. 20: 1769–1775.

Ruddiman, William. 2003. "The Anthropogenic Greenhouse Era Began Thousands of Years Ago." *Climate Change*, Vol. 61, No. 3: 261–293.

———. 2006. *Ploughs, Plagues, and Petroleum*. Princeton, NJ: Princeton University Press.

Sale, Kirkpatrick. 1991. *Dwellers of the Land*. Philadelphia, PA: New Society Publishers.

———. 1995. *Rebels Against the Future: The Luddites and Their War on Industrial Revolution*. Reading: Addison Wesley.

Sassen, Saskia. 2006. *Territory, Authority, Rights: From Medieval to Global Assemblages*. Princeton, NJ: Princeton University Press.

Schäfer, Wolf. 2007. "Knowledge and Nature: History as the Teacher of Life Revisited." *Nature and Culture*, Vol. 2, No. 1, Spring: 1–9.

Scheidel, Walter. 2017. *The Great Leveler: Violence and the History of Inequality from the Stone Age to the Twenty-First Century*. Princeton, NJ: Princeton University Press.

Schmidt, Alfred. 1972. *The Concept of Nature in Marx*. London: New Left Books.

Scobie, Robert, and Israel, Shel. 2016. *The Fourth Transformation: How Augmented Reality and Artificial Intelligence Change Everything*. New York: Patrick Brewster Press.

Sessions, George. 1985. "Western Process Metaphysics." In *Deep Ecology*, edited by Bill Devall and George Sessions, 236–243. Salt Lake City, UT: Peregrine Books.

——— (ed.). 1995. *Deep Ecology for the 21st Century*. Boston: Shambala.

Sinor, D. 1977. *Inner Asia and Its Contacts with Medieval Europe*. London: Variorum.

Smil, Vaclav. 2004. *China's Past, China's Future*. New York: Routledge.

———. 2008. *Energy in Nature and Society*. Boston: MIT Press.

———. 2014a. *Why America Is Not a New Rome*. Boston: MIT Press.

———. 2014b. *Making the Modern World: Materials and Dematerialization*. New York: Wiley.

———. 2016. *Still the Iron Age*. New York: Butterworth-Heinemann.

———. 2017. *Energy and Civilization: A History*. Boston: MIT Press.

Snodgrass, Anthony. 1971. *Archaic Greece*. Berkeley: University of California Press.

Snyder, Gary. 1974. *Turtle Island*. New York: New Directions Books.

———. 1980. *The Real Work*. New York: New Directions Books.

Spinoza, Benedictus de. 1941. *Ethics*. New York: E.P. Dutton.

Tegmark, Max. 2017. *Life 3.0*. New York: Knopf.

Thayer, Robert L. 2003a. *Life-Place*. Berkeley, CA: University of California Press.

———. 2003b. *Lifeplace: Bioregional Thought and Practice*. Berkeley: University of California Press.

Thompson, William R. 2009. "Structural Preludes to Systemic Transition." In *Systemic Transitions: Past, Present and Future*, edited by William R. Thompson, 55–74. New York: Palgrave.

Thoreau, Henry David. 1972. *The Maine Woods*. Princeton, NJ: Princeton University Press.

———. 1982. *Thoreau in the Woods*. New York: Farrar, Straus, Giroux.

Torrance, Robert (ed.). 1998. *Encompassing Nature: A Source Book on Nature and Culture Relations from Ancient Times to the Modern World*. Washington, DC: Centerpoint Press.

Totty, Michael. 2019. "The Worlds that AI Might Create." *Wall Street Journal*, October 13, 2019.

Trentmann, Frank. 2016. *Empire of Things*. New York: Harper Collins.

UNHCR. 1995. *Statistical Yearbook Populations of Concern to UNHCR: A Statistical Overview*. Geneva: UNHCR.

United Nations. 2003. *World Population Prospects 2002*. New York: Department of Economics and Social Affairs.

———. 2005. *World Economic and Social Survey 2004: International Migration*. New York: United Nations.

United Nations Conference on Environment and Development. 1992. *Earth Summit '92*. London: The Regency Press.

United Nations Environment Programme. 2002. *World Environment Outlook*. Vol. 3. Nairobi: UNEP.

United Nations Fund for Population Activities. 2007. *State of the World Population 2007*. New York: UNFPA.

Wallerstein, I. 1974, 1980. *The Modern World-System Vols 1–3*. New York: Academic Press.

———. 1988. "Kondratieff Up or Down." *Review*, Vol. 2, No. 4: 663–673.

———. 1991. "World System Versus World-Systems: A Critique." *Critique of Anthropology*, Vol. 11, No. 2: 189–194.

———. 1992. "The West, Capitalism, and the Modern World System." *Review*, Vol. 15, No. 4: 561–620.

Walters, C. C. 1974. *Monastic Archaeology in Egypt*. Warminster: Aris and Phillips.

WBGU. 2011. *Factsheet #3 Urbanization*. Berlin: WBGU Secretariat.

Westra, Laura. 1997. "Aristotelian Roots of Ecology." In *Environment and the Greeks*, edited by L. Westra and T. Robinson, 23–32. London: Rowman and Littlefield.

Whitaker, Mark D. 2006. *Ecological Revolution: The Political Origins of Environmental Degradation and the Environmental Origins of Axial Religions; China, Japan, Europe*. Koln, Germany: Lambert Academic Publishing.

White, Lyn, Jr. 1967. "The Historical Roots of Our Ecological Crisis." *Science*, No. 155: 1203–1207.

Wilson, Edward O. 2002. *The Future of Life*. New York: Knopf.

———. 2016. *Half-Earth*. New York: Knopf.

———. 2019. *Genesis*. New York: W.W. Norton.

Wilson, Ernest. 2004. *The Information Revolution and the Developing Countries*. Cambridge, MA: MIT Press.

Wladawsky-Berger, Irving. 2019. "How to Support Widespread Adoption of Artificial Intelligence." *Wall Street Journal*, October 4, 2019.

Wolfram, Herwig. 1988. *History of the Goths*. Berkeley: University of California Press.

———. 1990. *The Roman Empire and its Germanic Peoples*. Berkeley: University of California Press.

Workman, H. B. 1913. *The Evolution of the Monastic Ideal*. Boston: Beacon Press.

World Commission on Environment and Development. 1987. *Our Common Future*. Oxford: Oxford University Press.

World Economic Forum. 2006. *The Global Information Technology Report 2006–2007*. London: Palgrave-MacMillan.

World Resources Institute. 2005. *State of the World 2005*. Washington, DC: World Resources Institute.

Worldwatch. 2018. *The State of Consumption Today*. http://www.worldwatch.org/node/810.

Wright Mills, C. 1956. *The Power Elite*. Oxford: Oxford University Press.

Zakaria, Fareed. 2020. "The New China Scare." *Foreign Affairs*, Vol. 99, No. 1: 52–69.

Zezima, Katie. 2011. "As Framers' Markets Go Mainstream Some Feel Glut." *New York Times*, August 21, 2011.

Zook, M. A. 2000. "Internet Metrics: Using Hosts and Domain Counts to Map the Internet Globally." *Telecommunications Policy*, Vol. 24, No. 6/7: 4–8.

———. 2001. "Old Hierarchies or New Networks of Centrality?: The Global Geography of Internet Content Market." *American Behavioral Scientist*, Vol. 44, No. 10: 1679–1696.

Index

About the Author

Sing C. Chew is the founding editor of the interdisciplinary journal *Nature+Culture*, and professor emeritus at Humboldt State University. He is the author of a trilogy on the ecological history of the world: *World Ecological Degradation: Accumulation, Urbanization and Deforestation 3000 B.C.-A.D. 2000* (Vol. 1), *The Recurring Dark Ages: Ecological Stress, Climate Changes, and System Transformation* (Vol. 2), *Ecological Futures: What History Can Teach Us* (Vol. 3), and a history of Southeast Asia entitled *The Southeast Asia Connection: Trade and Polities in the Eurasian World Economy 500 BC-AD 500*.

www.ingramcontent.com/pod-product-compliance
Ingram Content Group UK Ltd.
Pitfield, Milton Keynes, MK11 3LW, UK
UKHW041844031026
15587UKWH00003B/993

9 781793 641502